Annual Meeting of the American Electrophoresis Society 2014 (AES)

Topical Conference at the 2014 AIChE Annual Meeting

Atlanta, Georgia, USA
16-21 November 2014

ISBN: 978-1-5108-1243-7

Printed from e-media with permission by:

Curran Associates, Inc.
57 Morehouse Lane
Red Hook, NY 12571

Some format issues inherent in the e-media version may also appear in this print version.

Printed by Curran Associates, Inc. (2015)

For permission requests, please contact AIChE
at the address below.

AIChE
120 Wall Street, FL 23
New York, NY 10005-4020

Phone: (800) 242-4363
Fax: (203) 775-5177

www.aiche.org

Additional copies of this publication are available from:

Curran Associates, Inc.
57 Morehouse Lane
Red Hook, NY 12571 USA
Phone: 845-758-0400
Fax: 845-758-2634
Email: curran@proceedings.com
Web: www.proceedings.com

TABLE OF CONTENTS

389532 Microchannel Electrophoresis for the Analysis of Amyloid Protein Oligomers

Abstract:

Amyloids are involved in a variety of diseases including Alzheimer's Disease and diabetes. A number of technical challenges exist for the quantitative analysis of the different sizes of amyloid proteins which are present during the early stages of aggregation, therefore requiring new technologies to be developed for enhanced protein separations. Microchannel electrophoretic techniques have emerged as powerful tools for the quantitative analysis of proteins. In this study, we investigated the ability of capillary and microchip electrophoresis to monitor the early stages of amyloid aggregation using amyloid beta (Aβ) and amylin as a model amyloid-forming proteins. Sample preparation effected the size of the initial starting material as well and the rate of protein aggregation. In addition, the use and selection of polymer coating material was found to significantly affect protein loss due to adsorption. Proteins were photo-crosslinked to explore the ability to resolve individual oligomeric species.

380529 Dielectrophoretic Preconcentration and Detection of Neuropeptides at Graphene-Modified Electrodes in a Nanochannel

Abstract:

Neuropeptides offer information on the chemistry, biology and pharmacology of the nervous system. However, their presence at picomolar levels within nanoliter-scale microdialysis samples poses analytical challenges due to interferences from other biomolecules at far higher levels. NPY is an especially significant biomarker, since it can be non-invasively collected from sweat, but its detection has been limited by poor sensitivity, long assay times and the inability to scale-down sample volumes. Herein, based on our prior work on dielectrophoretic trapping of proteins within nanochannels [1-3], we apply electrokinetic preconcentration of the neuropeptide onto patterned graphene-modified electrodes [4] in a nanochannel [5] by frequency-selective dielectrophoresis (DEP). While DEP has been extensively applied previously to sort micron-sized biological cells based on their electrophysiology, its application to separate nanoscale biomolecules in physiological media is limited by the cube-fold drop in trapping force with size of the polarized particle and an increase in disruptive electrothermal flow due to non-uniform Joule heating. We have overcome this limitation by using a nano-slit device design with sharp constrictions to enhance trapping forces, while reducing Joule heating due to enhanced heat dissipation at nanochannel interfaces. In this manner, through localized concentration enhancement of NPY on graphene-modified electrodes in a nanochannel that were optimized for high conductivity and fast adsorption kinetics, we demonstrate picomolar detection sensitivities within seconds from sub-nanoliter micro-dialysates. In follow-up work, we are exploring the coupling of this paradigm with aptamers for improving detection selectivity for electrochemical detection of neuropeptides within bio-fluid matrices including interfering electroactive peptides. Based on the ability of this methodology for detection sensitivities comparable to immunoassays, but with far more rapid assay times than immunoassays and with far less sample volumes, we envision its application within off-line detection from dialysates for measurement of neuropeptides at high spatial and temporal resolutions.

References

1. K.T. Liao, M. Tsegaye, V. Chaurey, C.F. Chou, N. Swami*. *Electrophoresis* (2012), 33, 1958-1966.
2. K. T. Liao, C. F. Chou, *J. Am. Chem. Soc.* (2012) *134*, 8742-8745.
3. V. Chaurey, A. Rohani, Y.-H. Su, W.V. Varhue, K.T. Liao, C. –F. Chou, N. S. Swami, *Electrophoresis* (2013) *34*, 1097-1104.
4. B. J. Sanghavi, S. Sitaula, M. H. Griep, S. P. Karna, M. F. Ali, N. S. Swami, *Anal. Chem.* (2013) *82*, 8158-8165.
5. B. Sanghavi, W. Varhue, J. Chavez, C.F. Chou, N. S. Swami*; *Anal. Chem.* (2014), DOI: 10.1021/ac500155g

384113 Human Fluid Sample Pretreatment for Biomarker Discovery

Abstract:

Human fluids – including plasma, serum, and cerebrospinal and pleural effusion fluids -- potentially are rich sources of protein biomarkers that can be used to indicate the presence or progression of disease or the response of disease to treatment regimens. Human fluids are enormously complex, with hundreds of thousands of different proteins, protein fragments and post-translational modifications that are present with concentrations that span ten orders of magnitude. Potential biomarkers are expected to be present at very low concentrations when compared with the abundant proteins, and the wide protein dynamic range far exceeds that of the analytical instrumentation used for the analysis, whether gel electrophoresis (such as DIGE) or bottom-up proteomic methods are used.

Nine protein biomarker panels have been introduced to date, but discovery of new ones has slowed considerably. We believe that one reason is that pretreatment methods applied to human fluid samples are inadequate. We review the extensive work described in the literature, and introduce new ones that we are developing at Sonata Biosciences. A particular emphasis is placed on treatment of the large volumes that are required to delve deeply into the proteome to uncover dilute protein biomarkers.

385590 Quantification of 2D Gel Western Blot Overlay Images

Abstract:

Previously, we have shown that epidermal growth factor receptor (EGFR) may be detected in human lung cancer samples with an anti-EGFR antibody using 2D gel western blotting. Furthermore, EGFR activated by tyrosine phosphorylation may be distinguished from inactive by colorizing and overlaying EGFR/pTyr western blot images from the same 2D gel. Visual comparison of overlay images shows the active EGFR is dramatically higher in some but not all tumors compared to the corresponding control lung tissue. But visual comparisons do not suffice when many samples are tested. In that case it is desirable to quantify images and express differences as a numerical value. The highly sensitive ECL/chemiluminescence used for EGFR is known to vary dramatically with time and amount of horseradish peroxidase (HRP) which oxidizes luminol to produce light. In this report we describe creation of an internal standard for chemiluminescence detection. The Lightning-Link HRP conjugation kit from Innova Biosciences was used to attach HRP to a purified protein, tropomyosin. In this case, no primary antibody is needed for western blotting, the HRP-tropomyosin reacts in the system to produce light exactly as the HRP-secondary antibody does in a normal western blot. Linearity of the HRP-tropomyosin (HRP-T) chemiluminescent signal within a 1D gel was demonstrated by loading varying amounts of HRP-T in different lanes and plotting band density versus concentration. Results are presented for quantification of EGFR protein spots detected by 2D western blotting in tumor and control samples using Progenesis SameSpots software with and without normalization to the internal HRP-T standard.

383749 Woven Fabric As a Low-Cost Microfluidic Platform for Tuned Electrophoretic Separations

Abstract:

Low-cost microfluidic devices have been widely researched as potential ways to making medical diagnostics and therapeutic monitoring more accessible to low income and low resource areas of the world. Current approaches to low cost device manufacture involve the assembly of materials such as chromatographic paper into microfluidic devices using multiple step approaches or microfabrication techniques such as UV-photolithography, which are expensive and difficult to scale up. In prior work, we demonstrated a textile weaving-based approach that makes use of this scalable and readily accessible technology to manufacture microfluidic devices in a single step. Liquid flow was tuned in these devices by manipulating the wetting characteristics and chemical treatment imparted to yarns in a seamless manner along the length of the fabric.

In recent work, we leveraged this tuning ability to manufacture fabric-based electrophoretic devices for the separation of mixtures of proteins. Electrophoresis is widely used for the pre-concentration, separation and assay of protein analytes in complex samples. We present the ability to tune separation resolution using the surface properties (wetting characteristics) and the packing density (number of yarns per unit area) of both woven and knit fabrics. These properties can be used to control sample dispersion. For instance, the use of hydrophilic yarns, such as cotton, in the separation channel allows a mixture of anionic dyes to migrate under the influence of an applied electric field, but disallows their separation. The converse is true for hydrophobic materials such as nylon and polyester, with greater packing densities resulting in well resolved separations and greater hydrophobicity resulting in faster separations. For dilute samples, a well known electrophoretic technique termed 'isotachophoresis' precedes separation. In this coupled method, a system of electrolytes of differing electrophoretic mobility are used in the buffer in order to first pre-concentrate (or stack) the proteins in the sample, following which the stacked sample is separated into sharp and well-resolved protein peaks. In gel electrophoresis, the separation event is triggered by a step change in pore-size and pH. We present the use of a seamless transition in yarn surface properties and packing density to perform protein sample pre-concentration followed by separation in a single fabric device. This forms the first instance of sample pre-concentration coupled with electrophoretic separation in a chromatographic platform, with potential applications in the detection of low abundance (rare) protein markers from clinical samples at the point-of-care or in low resource settings.

378358 Using Proteins As Resolution Probes to Quantify Gradient Insulator-Based Dielectrophoresis

Abstract:

Dielectrophoresis (DEP) is an ideal technique for separation and concentration of small volume samples. Dielectrophoresis has typically been applied to particles or cells, but recently smaller analytes, such as proteins, have been targeted. Concentration of proteins by dielectrophoresis has been demonstrated, with the first observations of protein DEP trapping about twenty years ago by Washizu et al. (IEEE Trans. Ind. Appl., 1994, 30, 835-843). The smallest proteins to be captured thus far are the proteins BSA (66 kDa) and streptavidin (60 kDa). Proteins of smaller molecular weights have been studied and manipulated using dielectrophoretic forces resulting in streaming (Electrophoresis, 2013, 34, 1085-1096). Here, we extend the range of explicit capture down to 14.3 kDa by capturing lysozyme from chicken egg white, trypsin inhibitor from soy, and other lower weight proteins.

Dielectrophoresis is carried out on a microfluidic device consisting of an insulating sawtooth-patterned microchannel such that an inhomogeneous electric field is induced in the channel when a DC potential is applied across the device. The gradient of dielectrophoretic forces in our device arises from the varying distances of each successive gate within the device: as the gates become narrower, the ratio of the dielectrophoretic force to the electrophoretic force increases. When the dielectrophoretic force is great enough to counteract all other forces a protein experiences within the channel, immobilization and concentration occurs.

One difficulty in attempting to quantitatively assess dielectrophoretic capture is that particles are on the same length-scale as the capture zone and local electric field gradient. When a number of particles are captured, they alter the local field and interact with one another. Smaller analytes such as proteins more faithfully reflect the forces present in the channel since they cause less of a disturbance to the electric field lines. Here we present capture of proteins and use these capture events to analyze the forces present within the channel. Electrokinetic mobilities and dielectrophoretic mobilities are calculated and compared with more traditional methods and numerical simulations.

390427 Characterizing the Dielectric Properties of Human Mesenchymal Stem Cells and the Effects of Charged Elastin-like Polypeptide Copolymer Treatment

Abstract:

Human mesenchymal stem cells (hMSCs) have three key properties that make them desirable for stem cell therapeutics: differentiation capacity, trophic activity, and ability to self-renew. However, current separation techniques are inefficient, time consuming, expensive and in some cases alter hMSCs cellular function. Dielectrophoresis (DEP) is a technique that uses alternating current (AC) electric fields to separate biological cells based on the dielectric properties of their membrane and cytoplasm. This work characterizes hMSCs dielectric signature and homogenizes its morphology. A quadrapole Ti-Au electrode microdevice was used to quantify the dielectric properties based on the cross-over frequency and DEP spectra of hMSCs from 0.01-35MHz in dextrose buffer solutions (0.03S/m and 0.10S/m). The membrane capacitance and permittivity were found to be 2.17pF and 4.53 in 0.03S/m and 1.70pF and 1.41 in 0.10S/m. Elastin-like polypeptide (ELP-) polyethyleneimine (PEI) copolymer was used to modify hMSCs morphology to spheroidal clusters. Results demonstrated that ELP-PEI treatment increased hMSCs membrane permittivity and shifted the cross-over frequency above 35MHz.

388633 Membrane Capacitance As a Label-Free Marker of Neural Stem Cell Fate

Abstract:

Understanding the factors regulating cell fate in the neural lineage is key for optimizing the use of neural stem and progenitor cells (NSPCs) for repair of the brain and spinal cord. However, little is known regarding the specific cell biological properties of neuron and astrocyte progenitors (NPs and APs) in this lineage. We found whole cell plasma membrane capacitance measured by dielectrophoresis is sufficient to distinguish and prospectively isolate NPs and APs from heterogeneous NSPCs without the use of labels.

These specific properties tied to cell fate may indicate the presence of cell membrane molecules involved in directing progenitors along neuronal or astrocytic fates. We hypothesized cell surface carbohydrates may contribute to fate-specific membrane capacitance. N-glycosylation controls cell surface expression and ligand binding of multiple classes of receptors affecting cell fate, and thus could regulate how cells respond to environmental cues.

We analyzed glycosylation enzymes by qRT-PCR and quantitated cell surface N-glycans by lectin binding and MALDI-TOF and found differences in glycosylation patterns between NPs and APs. Treatment of cells with specific agents to modify glycosylation pathways led to shifts in membrane capacitance. Furthermore, alteration of the glycosylation patterns on the cell surface was sufficient to change cell fate. Accurate patterns of glycosylation are critical for neural development in humans and rodents, highlighting the importance of glycosylation to NPs and APs in vivo. Our data indicate clear links among cell surface glycosylation, membrane capacitance, and cell fate, suggesting N-glycosylation is a critical regulator of cell fate in the neural lineage.

374921 Membrane Capacitance: A Biomarker for Tumourgenicity/Stem Cell-like in Human Oral Cancer Cells

Abstract:

Oral cancer is the 6th most common cancer in the world. Oral Squamous Cell Carcinoma (OSCC) accounts for around 90% of oral cancer. It is an aggressive cancer with very poor survival rate, which is in part due to tumour recurrence after drug treatment. Growing evidence has pointed to another potential contributing factor, which is the cellular "stemness" behaviour within the tumour itself, giving rise to heterogeneity and subpopulations of cancer cells with "stem-like" characteristics. These subpoplulations play a role in increasing tumorgenesis, resist anticancer treatments and permitting tumour recurrence. Despite controversy that surround these stem-like subpopulations, a lot of work has been carried out in this area in order to understand their physiology and their potential mechanisms that may allow better drug targeting and eradication.

Dielectrophoresis (DEP) has been used as a characterization tool for many years, by fitting the Clausius-Mossotti factor as a function of frequency to the data, a DEP spectrum (or "fingerprint") can be produced and the electrophysiological parameters can be extracted. These parameters include cytoplasmic conductivity (an indicator of ionic strength within the cytoplasm), effective membrane conductance (Geff)-indicative of how well the ions are conducting across the membrane, and effective membrane capacitance (Ceff)- indicative of membrane morphology (foldings and roughness).

The aim of this work was to use DEP to investigate the electrophysiological properties of OSCC cells with subpopulations of varying tumourgenic tendencies, in order to establish whether a "fingerprint" exists that can identify those subpopulations and provide a potential tool for the non-invasive, label-free and reliable isolations of these putative tumorgenic cells.

Using rapid adherence to collagen IV, subsets of cells with different tumorigenic abilities were isolated from a panel of OSCC and dysplastic cell lines, and their tumourgenicity was assessed using translational methods in mice. Furthermore, the membrane morphology study was also further supplemented with scanning electron microscopy and subsequent modulation with 4-methylumbelliferone (4-MU). We used one oral dysplastic and three OSCC-derived cell lines. Our results showed that the rapid adherent cells (RAC) to collagen IV, enriched for increased tumorigenic ability, had significantly higher effective membrane capacitance than middle (MAC) and late (LAC) adherent cells. SEM results showed a corresponding rough membrane surface that appeared to be very rich in protrusions (filopodia) relative to MAC and LAC. Subsequent treatment with 4-MU significantly reduced the effective membrane capacitance seen and in turn, a reduction of these membrane protrusions. To our knowledge, this is the first study that uses DEP to study OSSC cells with different tumorigenic abilities. It showed that these cells possessed different electrophysiological properties, and demonstrated the potential use of effective membrane capacitance as a biomarker of tumourgenicity and the stem-cell likeness of human oral cancer cells. These results suggest that DEP could potentially be used in the future for reliable, label-free isolation of putative tumorigenic cells.

Electrical tweezer for highly parallelized electro-rotation measurements over a wide frequency bandwidth for characterizing microbial subpopulations

A. Rohani,[a] W. Varhue,[a] Y.-H. Su,[a] and N. S. Swami[a*]
[a] Department of Electrical & Computer Engineering
University of Virginia, Charlottesville, VA 22904, USA
*Corresponding Author's E-mail: nswami@virginia.edu

Electrorotation (ROT) is a powerful tool for characterizing the dielectric properties of cells and bio-particles. However, its application has been somewhat limited by the need to mitigate disruptions to particle rotation by translation under positive dielectrophoresis (pDEP) and by frictional interactions with the substrate [1, 2]. While these disruptions may be overcome by implementing particle positioning schemes or field cages, these methods restrict the frequency bandwidth to the negative dielectrophoresis (nDEP) range and permit only single-particle measurements within a limited spatial extent of the device geometry away from field non- uniformities. Herein, we present an electrical tweezer methodology based on a sequence of electrical signals, composed of negative dielectrophoresis using 180-degree phase shifted fields for trapping and levitation of the particles, followed by 90-degree phase shifted fields over a wide frequency bandwidth for highly-parallelized electro-rotation measurements. Through field simulations of the rotating electrical field under this wave-sequence, we illustrate the enhanced spatial extent for electro-rotation measurements, with no limitations to frequency bandwidth. We apply this methodology to characterize subpopulations of *Cryptosporidium parvum* and *Clostridium difficile* with altered electrophysiology from the majority population [3] based on shifts in the electro-rotation spectra over the 0.05-40 MHz region [4]. Given the single particle sensitivity and the ability for highly parallelized electro-rotation measurements, we envision its application towards characterizing heterogeneous subpopulations of microbial and stem cells.

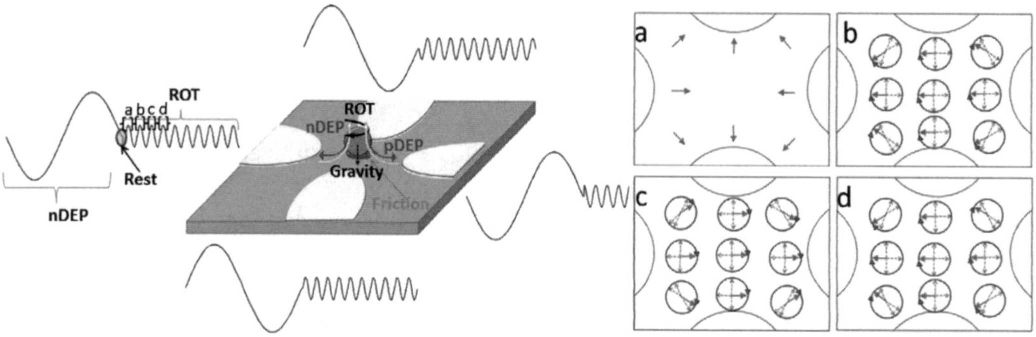

Figure 1: (left) Schematic of device geometry with the wave-sequence signal at neighboring electrodes, with simulations of the rotating electric field across the device geometry over 1 μs time periods at the four indicated time windows (a-d) under the wave-sequenced nDEP (4x) with ROT fields.

References

1. K. Maswiwat, M. Holtappels, and J. Gimsa, "On the field distribution in electrorotation chambers - Influence of electrode shape," *Electrochimica Acta,* vol. 51, pp. 5215-5220, Jul 15 2006
2. M. P. Hughes, S. Archer, and H. Morgan, "Mapping the electrorotational torque in planar microelectrodes," *Journal of Physics D-Applied Physics,* vol. 32, pp. 1548-1552, Jul 7 1999.
3. Y. H. Su, M. Tsegaye, W. Varhue, K. T. Liao, L. S. Abebe, J. A. Smith, R.L. Guerrant, N. S. Swami, "Quantitative dielectrophoretic tracking for characterization and separation of persistent subpopulations of Cryptosporidium parvum," *Analyst,* vol. 139, pp. 66-73, 2014.
4. "Electrical tweezer methodology for highly parallelized electro-rotation measurements over a wide frequency bandwidth", A. Rohani, W. Varhue, Y.-H. Su, N.S. Swami*; *Electrophoresis* (2014), DOI: 10.1002/elps.201400021

384621 Integrated Ion-Exchange Membrane Based Microfluidic Platform for Early Detection of Oral Cancer

Abstract:

Oral cancer belongs to a group of head and neck cancers that has shown a steep increase in its incidence in the last few years. According to The Oral Center Foundation, approximately 43,250 people in the US will be newly diagnosed with oral cancer, a cancer that already claims 7000 deaths annually. In addition to the excessive use of alcohol and tobacco, another major causative agent of the cancer is human papillomavirus serotype 16 (HPV16). HPV16 is found in approximately 40-80% patients. If diagnosed early, the chances of survival are very high (around 80-90%). However, this number decreases dramatically (5-year survival rate is less than 45%) if the tumor is revealed at a late stage, which is usually the case. Thus there is a pressing need for an early screening tool for detection of HPV-based oral cancer.

Recent studies have revealed that a panel of miRNAs contained in patient's saliva may be good oral cancer biomarkers. These miRNAs in saliva can either be free (mostly confined in exosomes) or come from cancer cells that have been shed off the oral tumor site. Saliva can be easily obtained from an individual during a preventive care visits in a dentist's office (liquid biopsy) without incurring any pain to the patient and it presents a good biological material for screening.

Our goal is to develop a screening platform for oral cancer that can process and analyze patient's saliva collected by oral rinses. The platform is designed as a microfluidic chip allowing automated sample processing and analysis based on the detection of miRNAs associated with oral cancer collected from saliva. The integrated platform can be divided into three independent units: (i) pretreatment unit for separation of miRNA molecules from the saliva samples, (ii) a unit to concentrate the loaded miRNA molecules and, (iii) ion exchange membrane sensor pre-functionalized with probes specific to oral cancer miRNA molecules. The miRNA detection is based on a so-called surface charge inversion phenomenon which occurs as a result of hybridization of negatively charged miRNA molecules to corresponding probes covalently attached on the positively charged ion exchange membrane. The hybridization is easily detected by measuring current-voltage characteristic of the membrane. The platform is capable of distinguishing two base-pair mismatches in a 22-base pairing segment of microRNAs associated with oral cancer. The limit of detection is 1 fM for short 27 base target molecules in a 15-minute assay. In addition to detection, we have successfully shown on-chip, efficient extraction and concentration of nucleic acid from a target sample into an automated single-chip diagnostic platform. The versatility and simplicity of this biosensor should enable point-of-care diagnostics of other cancer biomarkers and more broadly for other medical diagnostics applications.

390424 Portable Smartphone-Enabled DNA Analysis

Abstract:

The lack of affordable, rapid, and easy to use diagnostic technologies is one of the most critical issues confronting global public health. Expanded availability of such technologies will not only impact health care, but will be a key to enabling rapid identification of infectious agents in environments with little or no laboratory infrastructure. Here we describe our recent efforts to address these needs by developing an innovative smartphone-enabled DNA analysis system based on the polymerase chain reaction (PCR). Our approach offers an inherently simple design (similar in principle to a lava lamp) that enables PCR to be completed in 10-20 minutes, as compared to 1-2 hours using conventional instruments. Electrical power requirements are dramatically reduced by harnessing natural convection to actuate the reaction, allowing the entire system to be operated from a standard USB connection via solar battery packs. Instantaneous detection and analysis are enabled using an ordinary smartphone camera and dedicated app interface. This simple to operate format makes it possible to perform PCR-based assays in scenarios where a "yes-no" result is desired and an extensive laboratory infrastructure is lacking. The simple and inexpensive design make it possible to construct a complete DNA analysis system for approximately $20. The ability to deliver performance comparable to or surpassing that of current-generation systems while simultaneously providing an orders of magnitude reduction in cost has potential to greatly expand the use of PCR-based detection assays by moving them out of the laboratory and into settings where they are needed most.

365159 Selective Concentration and Separation of Colloidal Particles By Positive Reservoir-Based Dielectrophoresis (rDEP)

Abstract:

Reservoir-based dielectrophoresis (rDEP) is a recently developed technique by our group that exploits the inherent electric field gradients at a reservoir-microchannel junction for particle and cell manipulations. It is based on the induced negative dielectrophoretic motion at the junction and has been demonstrated to trap and isolate particles based on size or surface charge and as well cells based on viability inside the reservoir. In this talk we present our recent experimental and numerical results for particle concentration and separation through positive rDEP. This separation is implemented by re-suspending a particle mixture into low-conductivity de-ionized water, which is compared with that in an electrolyte solution that induces negative rDEP.

377142 Ising Lattices of Asymmetric Colloidal Dimers Under Electric Fields

Abstract:

Studies on colloidal particles have shed light in the physics of soft materials, including the crystals nucleation and growth, phase behavior, and glass formation. In particular, anisotropic particles can potentially mimic natural interactions between molecules, which are often shape-specific and orientation-dependent. Arguably, the simplest form of anisotropic particles beyond spheres is colloidal dimer where two lobes are grown or fused together. Here we report the directed-assembly of geometrically anisotropic colloidal dimers under the influence of AC electric fields. By combining the anisotropy in both particle geometry and field-induced electrostatic/hydrodynamic interactions, we show a variety of novel structures that closely resemble one- and two-dimensional Ising lattices, including small clusters, linear chains, square crystals, and frustrated triangular arrays. All of them are uniquely formed by alternating association between dimers with opposite orientations. To understand the underlying physics of the structural formation, we develop a theoretical model based on the energetic approach. Our model reveals that the overall pair interactions between asymmetric dimers depend sensitively on their relative orientations. Moreover, the potential between oppositely oriented dimers consists of a concave shoulder with two different length scales, which well explains the unique structures that we observe in experiments.

369306 Local Electrochemical Kinetics Allows for the Formulation of Physically Realistic Boundary Conditions for Fast Electrokinetic Applications

Abstract:

Microfluidic and nanofluidic applications such as a single nanopore sensing, AC electroosmosis or dielectrophoresis are often driven by high-frequency electric fields. Source and sensing electrodes may be in a direct contact with aqueous electrolytes, which is accompanied by an electron transfer through the electrolyte-electrode interface. Due to small characteristic dimensions of microfluidic and nanofluidic devices, electric double layers (EDLs) formed at the electrodes can overlap and, more importantly, it is difficult or impossible to define bulk concentrations and a reference electric potential. Further, fast perturbations caused by high-frequency fields can deviate EDLs far from the thermal equilibrium and all conclusions based on the Boltzmann distribution assumption can be violated. In the above mentioned cases, the Poisson-Nernst-Planck boundary value problem should be analyzed and solved. If the Faradaic interactions at the electrode-electrolyte interface cannot be neglected, boundary conditions containing the electron transfer via electrochemical reactions have to be formulated. The use of the classical Butler-Volmer kinetics or its Frumkin modification is limited because of the absence of an electrolyte bulk and fast changes of the electric field. For that reason, we developed an alternative kinetics scheme of n-electron electrochemical transfer that exploits only the local concentrations of electrochemical reactants at electrode-electrolyte interfaces. The scheme is based on the idea of the formation of electron donors and acceptors in the solid phase due to the presence of nonzero electric charge. These mediators enter electrochemical reactions and produce reduced and oxidized forms of the electrochemical reactants. The suggested kinetic scheme is analogical to the elementary kinetics widely used in heterogeneous catalysis. We show complete formulation of the boundary value problem with a reversible electrochemical interaction in a case study. Further, we suggest a qualitative structure of the electrode-electrolyte interface that takes into account the electric potential drop in a thin surface layer of the conductive solid phase, which corresponds to the presence of nonzero electric charge in the electrode. Finally, we prove that the suggested kinetic mechanism is in agreement with the classical Nernst theory.

Přibyl M. and Šnita D, Local kinetics and thermodynamics of rapid electrochemical reactions, Phys. Rev. E 89, 042403, 2014.

Červenka P., Hrdlička J., Přibyl M., and Šnita D, Kinetic mechanism for modeling of electrochemical reactions, Phys. Rev. E 85, 041505, 2012.

397554 Optoelectric Assembly and Manipulation of Beads in a Vertical Tower Configuration

Abstract:

A number of optical and optoelectrical techniques have been proposed for the trapping and manipulation of micro- to nanoscale colloidal particles. Rapid Electrokinetic Patterning (REP) is one such technique. It uses laser induced electrothermal flow and particle-electrode interactions to reversibly capture particles in a monolayer. Particle trapping in a monolayer is mainly governed by AC Electroosmosis, electrothermal flow, van der Waals attraction and electrostatic forces. Particles can be translated on the substrate by optically steering the laser or by moving the translation stage. In this work we show that by a careful selection of parameters, particles can be rapidly trapped in a vertical tower configuration instead of a monolayer. Moreover, this vertical tower can be translated and positioned at any desired place on the chip. We intend to discuss underlying physical mechanism and potential applications in biology.

387476 Electrokinetic Colloid and Micro-Vortex Dynamics in Heterogeneous Nano-Slot Devices

Abstract:

Understanding the electrokinetic interaction between colloidal particles and a nanochannel is of particular interest in the field of micro- and nano-fluidics and may play an important role in biomolecular detection. At relatively high-voltages, wherein concentration-polarization (CP) and electro-convective vortices emerge, the interaction complexity is further increased. Here we study the colloid dynamics in the vicinity of a heterogeneous nanoslot device wherein field-focusing effects and geometrically modulated interface significantly affect the CP and vortices due to both Dukhin's electro-osmosis of the 2nd kind and Rubinstein's electro-convective instability. In particular, trapping of colloids at the nanoslot entrance and their accumulation due field-focusing gradient effects within the diffusion layers is addressed.

389850 Size, Shell Material and Medium Conductivity Dependence on Dielectrophoretic Behaviors of Air Core, Chitosan/Poly-l-Lysine Shell Nanoparticles: Experimental Results

Abstract:

Dynamic (dis)assembly of biocompatible nanoparticles into 3D, packed structures would benefit drug delivery, films and diagnostics. Dielectrophoretic (DEP) microdevices can rapidly assemble, discriminate, and manipulate polarizable particles within non-uniform electric fields. DEP has primarily discerned micrometer particles; nanoparticles are less explored due to smaller forces that are only an order of magnitude or two larger than Brownian motion. In this work, we examine size and shell material dependencies via DEP of spherical core-shell nanoparticles (CSnp) into 3D particle-assemblies. Two types of CSnp shell materials (chitosan and poly-L-lysine) were custom synthesized around a phospholipid micelle core containing perfluorooctyl bromide oil. For DEP experiments, frequencies from 100kHz to 80MHz at fixed 5Vpp and medium conductivities of 10^{-5} and 10^{-3} S/m were tested. The DEP responses of ~220 and 340(or ~400) nm chitosan/poly-L-lysine CSnp were quantified via video intensity densitometry at the microdevice's quadrupole electrode center, translated into a time averaged relative DEP force, then compared to a spherical core-shell model. The core-shell particles experienced negative DEP across all frequencies, medium conductivities, shell materials and sizes tested. This is attributed to shell and core conductivity and permittivity values less than the medium. Frequency dependent DEP responses revealed a resonance peak between 55MHz-62MHz, which was size and shell material dependent. This data provides insight into organic shell material properties and is concurrently useful for dynamic assembly applications.

385908 Electrohydrodynamic Particle Structuring on a Drop Interface

Abstract:

We experimentally study the dynamics of colloidal particles adsorbed on a drop placed in a uniform DC electric field. The interplay of dielectrophoretic attraction and convection by the electrohydrodynamic flow gives rise to variety of particle assemblies. In weak electric fields, particles move towards the equator forming a "belt". As the field strength increases, the belt becomes unstable and breaks into a sequence of counter-rotating vortices of particles. In strong fields, the particles chain and may tip-stream from the drop poles. The diversity of particle patterns and drop shapes provides new ideas for directed assembly of particles at interfaces. The observed dynamics may also apply to other complex surfaces, such as domains on fluid bilayer membranes.

385162 The Effects of Charge Relaxation and Charge Convection on Nonlinear Electrohydrodynamic Drop Deformation

Abstract:

We quantify the nonlinear transient deformation of a weakly conductive ("leaky dielectric") drop under the influence of a uniform DC electric field. An axisymmetric boundary integral method, which accounts for surface charge convection and a finite relaxation time scale over which the interface charges, is implemented to calculate the transient droplet deformation. The simulations show that as the time scale for interfacial charging increases, a shape transition from prolate (parallel to the applied field) to oblate (normal to the applied field) occurs at intermediate times. Our work identifies the scaling that characterizes the magnitude and duration of this transition. Additionally, convection of charge towards the equator of the drop is shown to weaken the steady oblate deformation as the applied field is increased. The numerical results are then compared with an experimental system consisting of a silicone oil droplet suspended in a castor oil medium. Although agreement is observed for moderate electric field strengths, the experiments exhibit an onset of rotation and three-dimensional flow as higher fields are attained.

390840 A Deep Tertiary Minimum in the Particle/Electrode Interaction Energy in Oscillatory Fields

Abstract:

Application of an oscillatory electric field is known to alter the equilibrium separation distance between micron-scale colloidal particles and an adjacent electrode. This behavior is believed to be partially due to a lift force caused by electrohydrodynamic (EHD) flow generated around each particle, with previous work focused on identifying a single equilibrium "height" of the individual particles over the electrode. Here we report the existence of a pronounced bifurcation in the equilibrium particle height in response to low frequency electric fields. Optical and confocal microscopy observations reveal that application of a 50 - 300 Hz field induces some of the particles to rapidly move several particle diameters up from the electrode, while the others remain near the electrode. Statistics compiled from repeated trials demonstrate that the likelihood for a particle to move up follows a binomial distribution, indicating that the height bifurcation is random and does not result from membership in some distinct subpopulation of particles. The fraction of particles that move up increases with increased applied potential and decreased frequency. Taken together, the results provide evidence for the existence of a deep tertiary minimum in the electrode-particle interaction energy at a surprisingly large distance from the electrode.

Electrokinetic Biosensing at Liquid Interfaces by Fluidic Dielectrophoresis

Nicholas Mavrogiannis and Dr. Zachary Gagnon
Johns Hopkins University
Nmavrog1@jhu.edu

Development of low-cost, rapid, sensitive and portable biosensing systems are important for the detection and prevention of disease in developing countries, biowarfare/anti-terrorism applications, environmental monitoring, point-of-care diagnostic testing and for basic biological research. Currently, the most established commercially available and widespread assays for portable point of care detection and disease testing are paper-based dipstick and lateral flow test strips. These paper-based devices are often small, cheap and simple to operate. The last three decades in particular have seen an emergence in these assays in diagnostic settings for detection of pregnancy, HIV/AIDS, blood glucose, Influenza, urinary protein, cardiovascular disease, respiratory infections and blood chemistries. Such assays are widely available largely because they are inexpensive, lightweight, portable, are simple to operate, and a few platforms are capable of multiplexed detection for a small number of sample targets.

However, there is a critical need for sensitive, quantitative and multiplexed detection capabilities for point-of-care diagnostics and for the detection and prevention of disease in the developing world that cannot be satisfied by current state-of-the-art paper-based assays. For example, applications including the detection of cardiac and cancer biomarkers and biothreat applications require sensitive multiplexed detection of analytes in the nM and pM range, and cannot currently be satisfied with current inexpensive portable platforms due to their lack of sensitivity, quantitative capabilities and often unreliable performance.

In this talk, I demonstrate how to perform inexpensive label-free biomolecular detection at liquid interfaces using a newly discovered electrokinetic phenomenon known as fluidic dielectrophoresis (fDEP). Our electrokinetic approach involves exploiting the electrical mismatches between two aqueous liquid streams which are forced to flow side-by-side in a microfluidic T-channel. In this system, we engineer one fluid stream to have a higher conductivity relative to its neighbor, while the adjacent flow stream has a lower permittivity.

When a "low" frequency (< 1 MHz) alternating current (AC) electrical field is applied normal to this fluidic electrical interface the fluid stream with high conductivity displaces into the low conductive stream. Conversely, when a "high" frequency (20 MHz) AC electric field is applied the high permittivity stream deflects across the microfluidic channel. There is, however, a critical frequency sensitive to the electrical differences between each fluid phase – the fDEP crossover frequency – between these two events where no fluid deflection is observed, and the interface remains fixed when exposed to an external field, shown below.

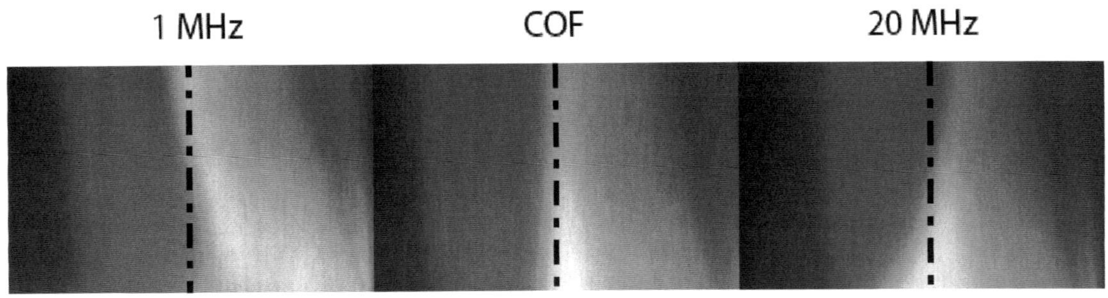

To perform biomolecular detection, we flow two streams side-by-side in our T-channel: one fluid stream with an analyte of choice and an adjacent stream with a specific receptor to the chosen target. We allow the two fluid streams to merge and measure the fDEP crossover at different axial positions down the resulting liquid interface in the microfluidic channel. We demonstrate that as biomolecular binding occurs, the electrochemical properties at the fluid interface become modified and can be readily detected by measuring the fDEP liquid crossover frequency.

Insulator-based micropipette dielectrophoretic trapping of particles

Stuart J. Williams, Daniel G. Allgeier

Department of Mechanical Engineering, University of Louisville, Louisville, KY 40292

We report on the progress of our research in pipette-based electrokinetic manipulation of particles. Through numerical simulations and preliminary experimentation, we explore the use of pulled glass pipettes to induce dielectrophoresis (DEP) and electrohydrodynamic flow. Typically, non-uniform electric fields for DEP are formed using micrometer metal electrodes fabricated with clean room processes. We will demonstrate that sufficient electric field gradients can be generated by applying an electric field through the tip of a tapered glass pipette (ID < 5 μm) which was fabricated without the use of a clean room or photolithography. Field gradients were generated similarly to insulator-based DEP and reservoir-based DEP by constricting the field with insulative features. Experimental trapping of 2 μm and 8 μm polystyrene particles is demonstrated at frequencies up to 130 kHz and voltages less than 35 V. Further, numerical simulations will demonstrate the magnitude of the theoretical DEP force, Joule heating, and electrothermal flow under these conditions.

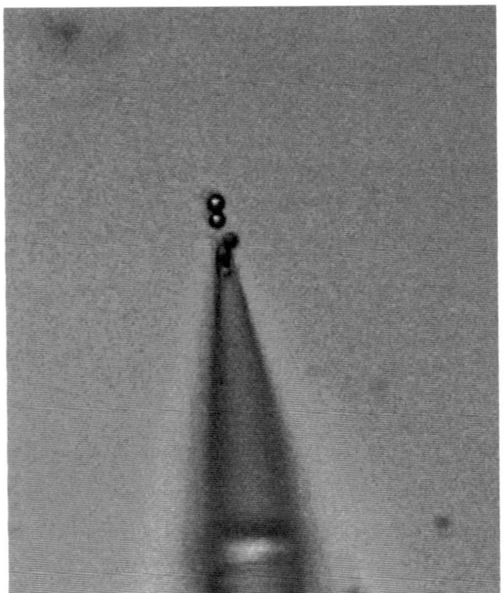

Figure 1. 8 μm polystyrene particles trapped near the tip of a pulled glass capillary at 120 kHz and 20 V.

397148 Optoelectric Trapping and Manipulation of Metal Nanoparticles

Abstract:

Noninvasive particle grippers are highly desired in nanomanufacturing and bio-microfluidics. Rapid Electrokinetic Patterning (REP) provides such functionality by combining AC electric field and laser induced heating on an electrode surface. In this technique, a colloidal solution containing nanoparticles is supplied between two planar transparent indium tin oxide (ITO) electrodes. Application of an AC electric field and the projection of a 1064 nm infrared beam on one of the electrode surfaces results in a toroidal electrothermal vortex which transports particles close to the electrode surface where the particles are reversibly trapped by electrode-particle interactions. We show that the metal nanoparticles such as gold and silver nanospheres (40-200 nm) and multi-wall carbon nanotubes can be trapped and manipulated using this method. We demonstrate application of REP induced aggregation in sensitive chemical detection with Surface Enhanced Raman Spectroscopy (SERS).

378170 Electric Field Directed Assembly of Anisotropic Colloids

Abstract:

Colloids with anisotropic interactions are good mimics of molecules. They could potentially assemble into more complex and unique structures which are impossible for isotropic spherical particles. Here we study the impact of geometric anisotropy on the assembly of asymmetric colloidal dimers under an external AC electric field. Experimental results clearly indicate that the assembly behaviors of asymmetric dimers strongly depend on their orientations. For example, the standing (lying) and standing (lying) dimers are well separated, while lying and standing dimers attract each other and assemble into a variety of chiral clusters, such as chiral trimers, tetramers, and pentamers. When all dimers stand on the substrate, dimers with opposite standing orientations attract each other and form one- and two- dimensional Ising lattice structures, including small clusters, linear chains, and squares. On the other hand, dimers with same standing orientations always repel each other. Our theoretical model based on electrostatic interactions agrees well with experimental results and provide further insights on electric-field assisted assembly of anisotropic particles.

The assembly of isotropic spherical colloids under external electric field is also investigated. Although the spherical particles are isotropic, "anisotropic" interactions between spherical particles can still be generated by applying an external AC electric field. This experiment is done within a previously unexplored experimental regime: low salt concentrations and low frequencies. The spherical particles assemble into well defined zig-zag oligomers from trimers, tetramers, to nonamers. Moreover, these oligomers can further connect and form two-dimensional non-close-packed networks. The assembly behaviors of spherical particles can be explained by the delicate balance among 1) offset dipolar attraction between top and bottom spheres, 2) dielectrophoretic attraction between top sphere and substrate, and 3) double layer repulsion between bottom spheres. These non-close-packed structures could be used as building blocks for making photonic crystals and plasmonic structures.

QUALITATIVE STUDY OF AC ELECTROOSMOTIC MICROPUMPS BY COMPUTATIONAL MODELING OF NON-EQUILIBRIUM ELECTROKINETICS

Abstract #388175

Matias Vazquez-Pinon[1,2], Victor H. Perez-Gonzalez[1], Sergio Camacho-Leon[1], Roberto C. Gallo-Villanueva[1], Marc J. Madou[1,2], Sergio O. Martinez-Chapa[1,2]*

[1]Sensors and Devices Group, School of Engineering and Sciences, Tecnologico de Monterrey, Mexico
[2]Department of Mechanical and Aerospace Engineering, University of California, Irvine, CA, USA
**Corresponding author: smart@itesm.mx*

ABSTRACT

In this work, qualitative results on computational modeling of three different electrode geometries employed to pump fluids in periodic sections of microchannels are presented. For this purpose, the non-equilibrium approach, which comprises the Poisson, Nernst-Plank, and Navier-Stokes equations, was solved through the finite element method. The modeled sections, which are 30 μm wide and 10 μm tall, are divided into quarters for an even distribution of blocks among different geometries. In this fashion, the outer quarters are gaps, while the two inner quarters are locations to fit the different electrode geometries. Our simulations show meaningful flow variations between the three types of electrodes, which are of up to three orders of magnitude. Through computational modeling, it is qualitatively assessed the difference between three electrode geometries, which is in good accordance with theory and experimental data that has been previously reported by other researchers.

INTRODUCTION

Electroosmotic flow (EOF) is the electrokinetic phenomenon related to the movement of fluid caused by the stimulation of mobile electric charges suspended in a solution. By means of viscous forces, the movement of electric charges in the form of ions drag fluid molecules, thus forging a velocity field along the solution in the direction of the moving charges. This movement is drawn by the electric double layer (EDL) formation that arises at the interface between a solid surface and the solution containing the free charges. The EDL is created due to electrostatic interactions between ions forming the outward layer of the solid and the neighboring mobile counter-ions in the fluid. The EOF is thus strongly related to the characteristics of the EDL such as the Debye length and electric field distribution. Figure 1 depicts the constitution of the EDL.

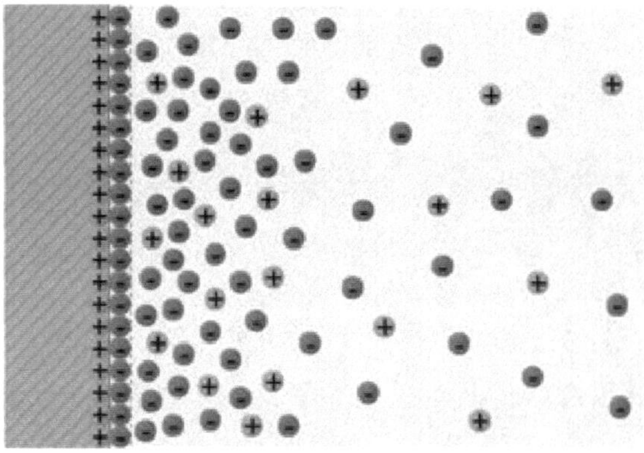

Figure 1. Constitution of the EDL.

To take advantage of this phenomenon, the EDL characteristics have to be intentionally controlled. This is accomplished by modifying its thickness by impelling an electric field into the bulk fluid by means of an external electric potential through conductive media. Typically, metallic electrodes are used to attain this task [1], but recently other materials have showed good results, such as glassy carbon [2], polysilicon [3], gel [4], and liquid metal electrodes [5]. The shape and size of the interface between the conductive structure and the electrolyte play an important role on the control of the EDL. With the correct design and implementation of the solid/liquid interfaces, EOF can be effectively employed to perform specific tasks in Lab-on-a-Chip systems such as pumping and mixing.

For this purpose, DC and AC potentials can be employed to induce the electric field across the electrolyte. In DC electroosmostic flow (DC-EOF), high potentials (~500 V) are longitudinally applied across the fluid. In response to this, an electric field is induced thus impelling the free charges to move along the field direction. Figure 2a shows the typical DC-EOF setup.

Alternatively, in AC electroosmotic flow (AC-EOF), low potentials are required (~1-20 V) since the electrodes are placed in the channel walls and commonly close enough together, and alternating signals are used to establish the flow direction. Figure 2b depicts the typical setup for AC electroosmotic flow. With this setup, the alternating field generates eddies that propels the bulk fluid with a net flow across the microchannel. Both DC- and AC-EOF are used to move fluids in the micro- and nano-scales, but the later has showed to be easier to integrate into microfluidic systems because it requires lower operating voltages and electrodes are easier to pattern on subtracts by means of photolithography. Additionally, AC-driven micropumps are straightforward to change from a fluid mixing mode to fluid transport mode, or to adjust the flow magnitude or to invert the flow direction.

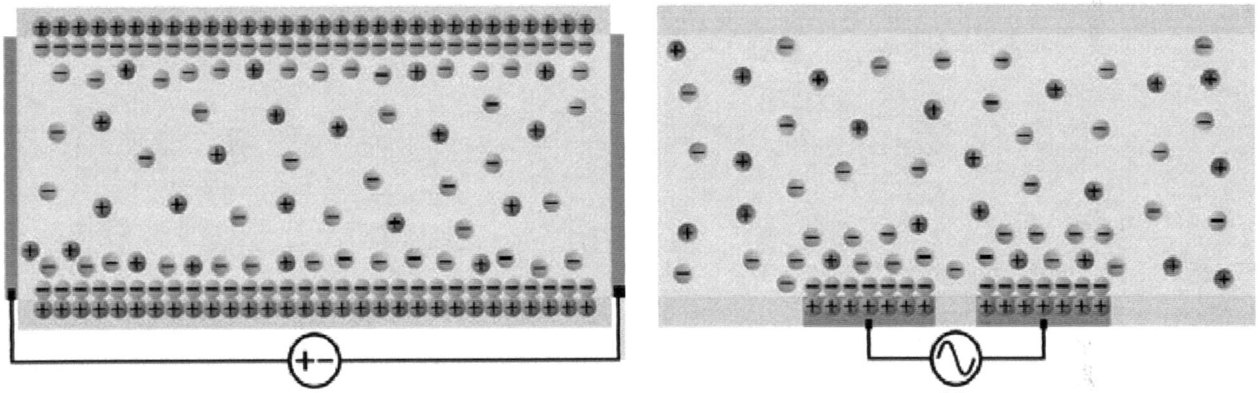
a) b)

Figure 2. a) DC Electroosmotic Flow and b) AC Electroosmotic Flow (EDL in walls omitted in this figure).

AC-EOF offers certain advantages over other approaches for microfluidic pumping and mixing. The foremost is the possibility of driving fluids without the need of mobile components within the microchannel. With this, some desirable benefits can be emphasized such as low cost and simplicity of fabrication, possibility of integration with other microdevices, and the ability to drive highly programmable flows at low electric potentials, thus low power consumption [6].

This work presents a qualitative study and comparison of three different micropumps based on AC-EOF, so called AC Electroosmotic micropumps (AC-EOM). These devices are capable of reaching flows in the range of commercially available pumps for microfluidic/nanofluidic applications. AC-EOMs are powered by an array of electrodes with characteristic geometries. The three most common arrays (asymmetric coplanar, traveling wave, and 3D or non-planar arrays) were modeled.

For this study, Finite Element analysis was employed to characterize the three mentioned arrays under similar conditions through the computational software COMSOL Multiphysics. The mathematical model, consisting on the Poisson-Nernst-Planck-Navier-Stokes equations, has been previously applied for modeling both, micropumps [7, 8] and mixers [9], but no comparison between the mentioned electrode geometries for AC-EOMs was found reported. In this work, the numerical model was implemented to compare the three different electrode geometries and their effect on the pumping performance in microchannels. Operational requirements and qualitative results on the net flow for the three cases are described.

NUMERICAL MODEL

The non-equilibrium approach comprises the Poisson equation to describe the distribution of the electric field strength, the Nernst-Planck equation to denote the sum of the diffusive, convective, and electro-migration flux intensities of the ionic components, and the Navier-Stokes and continuity equation to define the velocity of the fluid and pressure [8, 10]. The electric field across the bulk fluid is defined as

$$\mathbf{E} = -\nabla \varphi \tag{1}$$

where φ is the differential potential between the electrodes. The Poisson equation relates the electric field and the volumetric electric charge density across the electrolyte:

$$\nabla \cdot \left(-\varepsilon \nabla \varphi\right) = \rho_e \tag{2}$$

Here, $\varepsilon = \varepsilon_0 \varepsilon_r$ is the relative permittivity of the medium, ε_0 is the permittivity of vacuum and ε_r is the dielectric constant of the medium. The electric charge density, ρ_e, for a symmetric electrolyte (1:1), is described as

$$\rho_e = F\left(c^+ - c^-\right), \tag{3}$$

where $F = 9.6485$ C/mol is the Faraday's constant. Positive (c^+) and negative (c^-) ionic concentrations are determined by the molar balances:

$$\frac{\partial c_i}{\partial t} + \mathbf{u} \cdot \nabla c_i = -\nabla \cdot \mathbf{J}_i + r_i. \tag{4}$$

Subscript i represents the two species, positive and negative, $\mathbf{u}$ is the velocity field $\mathbf{J}_i$ and r_i are the ionic fluxes and total sources in all chemical reactions for the i_{th} component, respectively. In this work, no chemical reactions were considered, i.e. $r_{+,-} = 0$. The ionic flux is given by the Nernst-Planck equation for convection, diffusion, and electromigration components (first, second, and third terms at the right side of eq. (5), respectively):

$$\mathbf{J}_i = \mathbf{u} c_i - D_i \nabla c_i - \frac{z_i D_i F c_i}{RT} \nabla \varphi \tag{5}$$

D_i is the diffusion coefficient and z_i the ion valence for both species, $R = 8.314$ JK/mol is the gas constant, and T is the room temperature. The ion mobility is determined by means of the Nernst-Einstein relation:

$$U_{m,i} = \frac{D_i}{RT} \tag{6}$$

Finally, the Navier-Stokes equation was employed for the conservation of momentum and the continuity equation to describe the conservation of mass. In this study, incompressible, Newtonian fluids were considered for all cases.

$$\rho\left(\frac{\partial \mathbf{u}}{\partial t} + (\mathbf{u}\cdot\nabla)\mathbf{u}\right) = -\nabla p + \mu\nabla^2\mathbf{u} + \rho_e\nabla\varphi \tag{7}$$

$$\nabla\cdot\mathbf{u} = 0 \tag{8}$$

where $\rho = 1000\ \text{kg/m}^3$ is the fluid density, p is the pressure and $\mu = 1\times10^{-3}\ \text{Pa}\cdot\text{s}$ is the dynamic viscosity of the fluid. In eq. (7), the last term, $\rho_e\nabla\varphi$, is the Coulombic force acting on the ions suspended in the solution, which is defined for an area with a height of 1 μm above the electrode lengths. Initial values for the electric potential, velocity field, and pressure are zero. Initial concentration for both species is $C_0 = 10\ \text{mol/l}$. Boundary conditions for all solid-liquid interfaces, including both, electrodes and channel wall surfaces, are defined as non-slip, i.e. $\mathbf{u} = 0$, while the insulating boundary condition, $\mathbf{n}\cdot\nabla\varphi = 0$, was established only for solid wall-liquid interfaces.

MODELING PROCEDURE

The geometries of the three electrode arrays are depicted in Figure 3. Electrode These electrodes are defined as boundary conditions, thus no resistive losses are considered. For a precise comparison on the performance of the micropumps, the microchannel dimensions were standardized to a transversal section of 30 μm by 10 μm, and variations were only made on the electrode geometries. Inside the microchannel, the three electrode sets were fit into a total length of 15 μm, which is distributed in the two inner quarters of the total channel width. For simplicity, the gravity effects are neglected from the results and the electrolyte is defined as an incompressible Newtonian fluid with constant relative permittivity, viscosity, density, and ionic diffusivities.

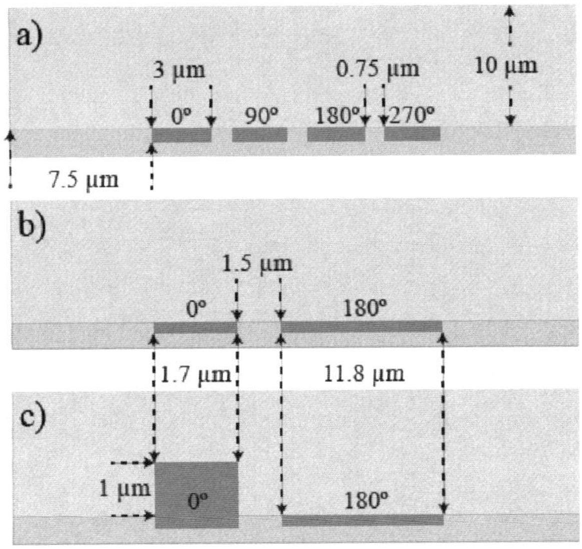

Figure 3. Electrode arrays: a) traveling wave, b) asymmetric coplanar, and c) 3D or non-planar.

For the traveling wave array, a set of four symmetric electrodes was used and four different phases where applied to the electrodes, $\varphi = A\sin(2\pi ft + \theta)$, where θ takes value of $0, \pi/2$, $3\pi/2$ and 2π for each subsequent electrode, as shown in Figure 3. For the cases of the asymmetric co-planar and 3D array, the geometries share the same electrodes and gaps lengths, and the only difference is the height of the short electrode, which is 1 μm for the 3D array. Furthermore, similar signals were applied to the electrodes with similar lengths in both cases. For the three cases of electrode arrays, a sinusoidal signal of 1V amplitude with a frequency of 1000 Hz was used as stimulation signal.

The structure of each micropump was modeled as a bi-dimensional geometry. The non-equilibrium approach was implemented using three interfaces: the *Electric Currents* branch for the definition of the electric field distribution, the *Transport of Diluted Species* branch to solve the Nernst-Planck equation and the *Creeping Flow* branch, which solves for the Navier-Stokes equations. Pertinent boundary conditions were established at liquid-solid interfaces and open boundaries were defined at the fluid inlet/outlet boundaries.

RESULTS

With the conditions described, vortexes are observed near the electrodes edges, which have different sizes due to the difference in the electrodes geometries: large vortexes above the large electrode and small ones above the small electrode. Due to this effect, a net flow is propelled on the bulk fluid with a net flow across the microchannel.

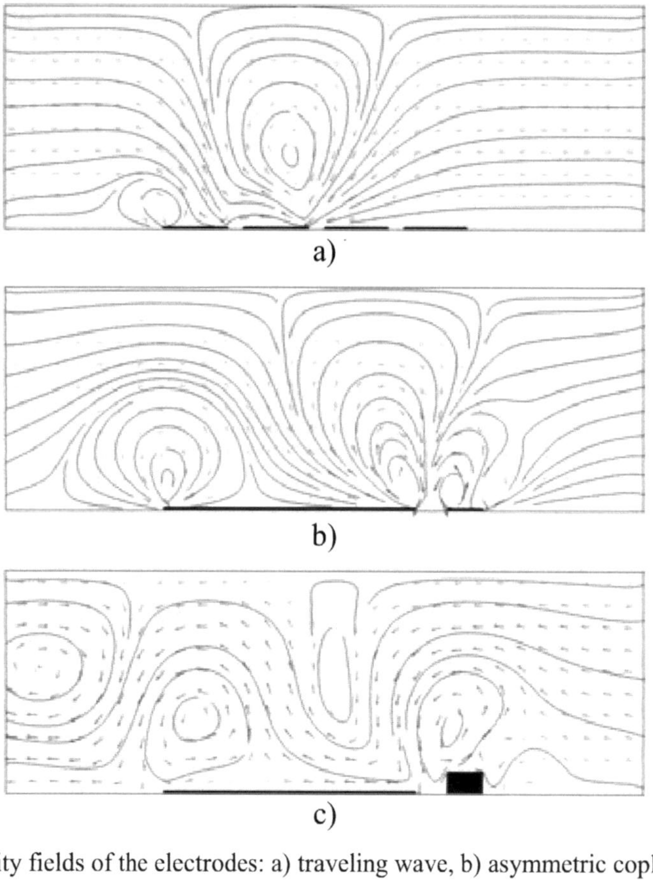

Figure 4. Velocity fields of the electrodes: a) traveling wave, b) asymmetric coplanar, and c) 3D.

There is a considerable increase in flow for the 3D array because the elevated electrode is blocking the back-flow generated while the electrode has a positive potential. From Figure 5, flow variations are in one order of magnitude.

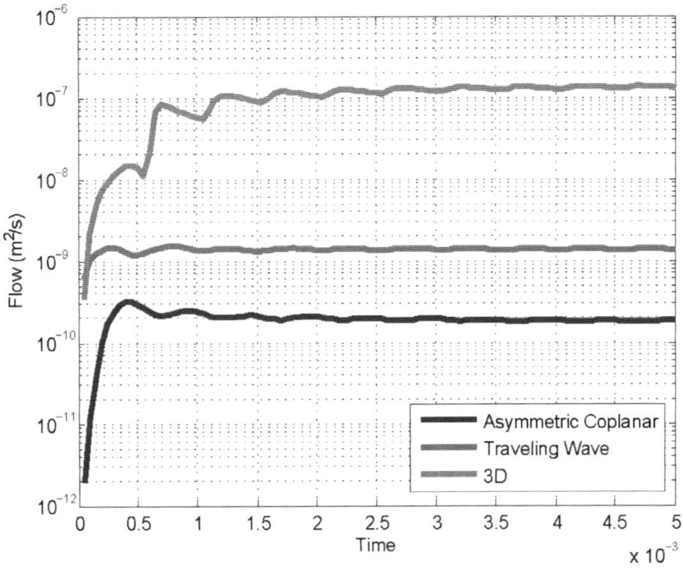

Figure 5. Comparison of the flow of the three geometries

For the same voltage signal in amplitude and frequency and the same fluid properties, the asymmetric co-planar array showed the lowest time-averaged flow of the three arrays. Flow in one order of magnitude higher than the asymmetric coplanar array was obtained with the traveling wave array, and the 3D array achieved a considerably higher flow than the traveling wave array and the asymmetric co-planar array (almost two orders of magnitude).

CONCLUSIONS

This work presented a qualitative study of three AC electroosmotic micropumps solving the Poisson-Nernst-Planck-Navier-Stokes set of equations by using the finite element method. The lowest fluid velocity is achieved by the asymmetric co-planar electrode array; the traveling wave electrode array achieves one order of magnitude higher than the asymmetric co-planar array, and the 3D electrode array is the configuration with the highest fluid flow. Both, the asymmetric co-planar array and the 3D array of electrodes generate a small back-flow that reduces the overall performance of the micropump, while the traveling wave array of electrodes preserve a positive flow along all the applied signal cycles.

Currently, our group is working on prototyping AC-EOMs with carbon and metal electrodes for planar and non-planar structures. We expect to assess the accuracy of the numerical models presented in this work by comparing the obtained results with experimental data and to propose an optimal design protocol that allows to fabricate AC-EOMs with performance specific for application needs.

REFERENCES

1. Green, N.G., et al., *Fluid flow induced by nonuniform ac electric fields in electrolytes on microelectrodes. I. Experimental measurements.* Phys Rev E Stat Phys Plasmas Fluids Relat Interdiscip Topics, 2000. **61**(4 Pt B): p. 4011-8.

2. Rouabah, H.A., et al., *Design and fabrication of an ac-electro-osmosis micropump with 3D high-aspect-ratio electrodes using only SU-8.* Journal of Micromechanics and Microengineering, 2011. **21**(3): p. 035018.
3. Guo, X., et al., *A study on three-dimensional electrode arrays fabricated by PolyMUMPs® for AC electro-osmotic pumping.* Microelectronic Engineering, 2011. **88**(10): p. 3113-3118.
4. Seibel, K., et al., *A programmable planar electroosmotic micropump for lab-on-a-chip applications.* Journal of Micromechanics and Microengineering, 2008. **18**(2).
5. Gao, M. and L. Gui, *A handy liquid metal based electroosmotic flow pump.* Lab on a Chip, 2014. **14**(11): p. 1866-1872.
6. Chen, J.S., et al., *Effects of Electrode Array Dimensions on the Pumping Performance of Traveling-Wave Electro-Osmotic Micro-Pumps.* Digest Journal of Nanomaterials and Biostructures, 2011. **6**(2): p. 391-402.
7. M. Přibyl, et al., *Non-Equilibrium Modeling of AC Electroosmosis in Microfluidic Channels - Parametrical Studies* -. International Journal of Plasma Environmental Science and Technology, 2009.
8. Hrdlička, J., et al., *Mathematical modeling of AC electroosmosis in microfluidic and nanofluidic chips using equilibrium and non-equilibrium approaches.* Journal of Applied Electrochemistry, 2009. **40**(5): p. 967-980.
9. Kim, D., et al., *Non-equilibrium electrokinetic micro/nano fluidic mixer.* Lab on a chip, 2008. **8**(4): p. 625-8.
10. Cervenka, P., et al., *Toward High Net Velocities in AC Electroosmotic Micropumps Based on Asymmetric Coplanar Electrodes.* IEEE Transactions on Industry Applications, 2010. **46**(5): p. 1679-1691.

Combination of vertical and planar electrodes for electrorotation and cell positioning in microchambers

Samuel C. Kilchenmann, Fabio M. Spiga, Carlotta Guiducci

Particle positioning and uniform electrical fields are enabling features for electrorotation experiments. Different approaches have been proposed so far to position particles inside a chamber. Schnelle et al. introduced the use of optical tweezers [1] while Han et al. showed that by modulation of two electrical signals particles can be rotated while they are trapped by dielectrophoresis [2]. In both cases, additional equipment is required, thus increasing both cost and complexity of the setup. For what concerns the homogeneity of the electric field, vertical electrodes would bring clear advantages with respect to planar electrodes[3]. Here we propose a microfabrication process for an electrorotation chamber consisting of vertical electrodes with planar extensions at their base, to achieve particle rotation and cell positioning with only four electrodes.

3D electrodes have the advantage that the electrical potential can be applied over the full channel height. Therefore, the resulting electric field and the rotational torque will be homogenous, with the torque independent from the z-position [3]. In the absence of dielectrophoretic forces in the direction orthogonal to the chip surface, the cell would only be subjected to the torque and to the gravitational force which might cause sedimentation in the time frame of the experiments. For this reason, we propose to include planar electrodes, placed at the base of the vertical electrodes, extending toward the center of the rotation chamber. These features would serve to induce levitation forces on the particles and therefore enable the control of its z-position. We performed finite element simulations of the structures represented in figure 1a, featuring parabolic electrodes of 50 μm height and 60 μm facing-electrode spacing, in order to study the electric field generated. Planar electrodes extend from the 3D structures by 10 μm. We found that the resulting DEP forces in z-direction mainly affect the bottom part of the rotation chamber (Fig. 1b). Furthermore, changes in the rotational torque along the z-axis are minimal as there is less than 1% difference between the top and the bottom of the chamber (Fig. 1c). By making use of negative DEP forces particles will automatically be placed in the center of the microchamber, while the height of the particles can be adjusted with the amplitude of the applied voltage, making use of the DEP levitation principle [4] (Fig. 1d). With such electrodes, it is thus possible to achieve particle trapping in the center of the electrorotation chamber without the need for additional signals and electrodes.

[1] T. Schnelle, T. Muller, C. Reichle, and G. Fuhr, "Combined dielectrophoretic field cages and laser tweezers for electrorotation," *Appl. Phys. B-Lasers Opt.*, vol. 70, no. 2, pp. 267–274, 2000.

[2] S.-I. Han, Y.-D. Joo, and K.-H. Han, "An electrorotation technique for measuring the dielectric properties of cells with simultaneous use of negative quadrupolar dielectrophoresis and electrorotation.," *Analyst*, vol. 138, no. 5, pp. 1529–37, Mar. 2013.

[3] S. C. Kilchenmann, F. M. Spiga, and C. Guiducci, "3D electrodes in microfluidic channels for electrorotation experiments," *Electrophoresis*, vol. submitted.

[4] P. R. C. Gascoyne and F. F. Becker, "Dielectrophoretic manipulation of particles," *IEEE Trans. Ind. Appl.*, vol. 33, no. 3, pp. 660–669, 1997.

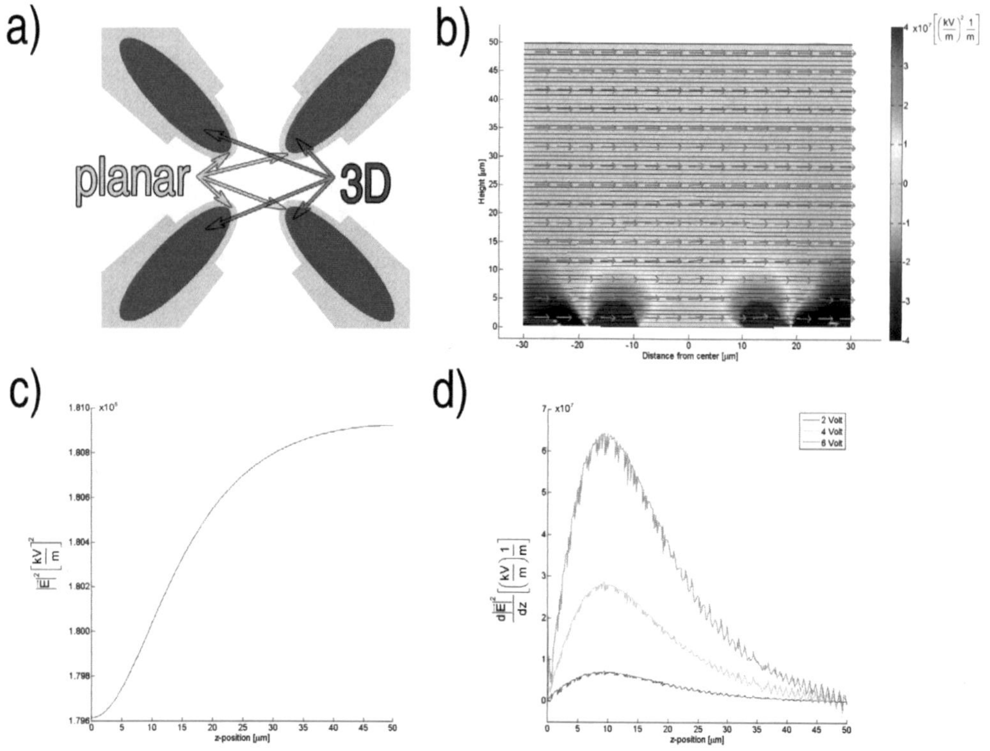

Fig. 1: a) top view of the extruded 3D electrodes (green) and their planar features (blue) b) simulation of DEP forces, the color indicates the z-component of the gradient of the electric field square (legend on the right), while the arrows indicate the direction of the DEP forces c) electric field square along the chamber centerline d) z-component of the gradient of the electric field square along the chamber centerline for different applied voltages.

383670 Brownian Dynamics Simulations of Electrophoretic DNA Separation in a Post Array By Pulsed Electric Field

Abstract:

Electrophoretic separation of DNA through a post array has been heavily investigated in both experiments and simulations. However, the efficiency of the post array for separating long DNA has been found decreased rapidly with increasing Peclet number (or electric field). The loss of resolution power is largely due to the onset of channeling phenomenon that DNA move through the hypothetical "channel" between posts without collisions with the posts. To improve the efficiency of separation at high Peclet number, we propose to use a pulsed electric field to replace the traditionally adopted constant electric field. When the electric field is turned off, DNA relax and diffuse, resulting in higher probability of collision with posts as the field is turned on. To test our idea, we have used Brownian dynamics simulations and computational fluid dynamics to simulate lambda-DNA(48.5 kbp) and T4GT7-DNA (166 kbp) under pulsed electric field through a hexagonal post array. It is found that the pulsed electric field does improve the resolving power of the post array at higher electric field due to the enhancement of the DNA-post collision.

384708 Experimental Study of DNA Electrophoresis in Post Array with Pulsed Electric Field

Abstract:

Microchannels with post array have been shown capable of separating long DNA in relatively short period of time than the conventional gel. However, the efficiency of the post array has been found decreased rapidly with increasing Peclet number (or electric field) due to the onset of the "channeling" phenomenon that DNA move through the hypothetical "channel" without collisions with the posts. In this study, we experimentally test the idea that using pulsed electric field with post array can attenuate the channeling phenomenon and regain the separation efficiency at high electric field. We have fabricated microchannels with array of posts of 1 micron in diameter on fused silica using standard photolithography and inductively coupled plasma (ICP) etching. The separation efficiency as a function of Pe is determined by measuring the electrophoretic mobility and the dispersion coefficient of λ-(48.5kbp) and T4(165.6kbp) DNA. The effect of the frequency and extension of the electric field pulse on the efficiency of DNA separation is also investigated.

390509 Frequency Sweep Rate Dependence on the Dielectrophoretic Response of Polystyrene Beads and Red Blood Cells

Abstract:

Dielectorphoresis (DEP) uses alternating current (AC) electric fields to polarize biological particles with experiments usually completed at static (fixed) frequencies. Developing DEP response spectra for biological particles from static frequency experiments is laborious, yet essential in determining the differences in dielectric properties. This work completed DEP experiments at multiple frequencies within a single experiment by sweeping the frequency with time to determine the DEP response spectra from less experiments. The optimal frequencey sweep rate was established using homogeneous 6.08μm polystyrene (PS) beads as a model system. This frequency sweep rate technique was extended to ~7μm red blood cells (RBC), to verify its applicability. A microfluidic device with Ti-Au quadrapole electrodes was used to subject the PS beads and RBCs to 10Vpp AC electric fields at frequencies between 0.010-2.0MHz with the sweeping rates ranging from 0.00080 to 0.17 MHz/s. The PS beads displayed only negative DEP behavior over these frequencies with the sweep rate influencing the response time. Faster sweep rates showed a lag in the DEP behavior for the PS beads as compared to the static frequency DEP responses, attributed to the dielectric relaxation time. This establishes the experimental parameter that the frequency sweep rate must be slower than the particle polarization time to achieve incremental polarization. Frequency sweep rates close to 0.00080 MHz/s yielded DEP behaviors consistent with static frequency DEP response for PS beads and RBCs.

Improving dielectrophoretic manipulation efficiency employing pitted and bumpy interdigitated electrodes

*Victor H. Perez-Gonzalez,[a] Vinh Ho,[b] Rene Celis-Cordova,[a] Lawrence Kulinsky,[b] Marc J. Madou,[a,b,c] and Sergio O. Martinez-Chapa,[a]**

a Department of Electrical and Computer Engineering, Tecnológico de Monterrey, Nuevo León, 64849, México.
b Department of Mechanical and Aerospace Engineering, University of California Irvine, California, 92617, United States.
c Department of Biomedical Engineering, University of California Irvine, California, 92617, United States.

*Correspondence author. Email: smart@itesm.mx

One of the most attractive features of Lab on a Chip (LOC) devices is portability. In this context, portability means being able to run an experiment without the need to use external lab equipment, such as bench-top function generators, to electrically stimulate the devices. Dielectrophoresis (DEP) is an electrokinetic technique that allows manipulating (concentrating, separating, filtering, characterizing, or displacing) microparticles in the presence of a spatially non-uniform electric field according to their dielectric properties. Moreover, DEP has great potential for miniaturization, which goes in line with the concept of portability in LOC devices.

There are two main research lines in dielectrophoretic theory, electrode-based DEP and insulator-based DEP (iDEP). Electrode-based DEP is commonly associated with the development of planar 2D electrodes to generate the required non-uniform electric field within the microfluidic channel. Although several different electrode topologies (e.g., interdigitated, castellated, or polynomial) have shown to enable microparticle manipulation through DEP employing low-voltage stimulation signals, many scientific reports have shown that 2D electrode design only affect a small portion of the complete sample under analysis. This means that only particles flowing in close vicinity to the electrodes respond to DEP, leaving the rest of particles in the bulk unaffected. In contrast to electrode-based DEP, iDEP permits affecting the complete volume of sample through the fabrication of 3D insulating obstacles that distort an electric field generated throughout the microfluidic channel. Nonetheless, iDEP has an important drawback. It requires very large stimulation voltages (around 1000V) for DEP to have a significant effect over the particles. This drawback makes it very difficult to integrate with microelectronic stimulation systems for the development of a fully-integrated LOC device.

Although CarbonDEP and PPyDEP have emerged in recent years as promising techniques for the development of 3D electrode-based DEP devices (overcoming the limitations of both, conventional electrode-based DEP and iDEP), the manufacturing processes involved are more complex than those required for conventional electrode-based DEP. Therefore, this work presents a novel and simple approach for the development of enhanced electrode-based DEP devices. Interdigitated electrodes with pitted or with bumpy surfaces are presented herein. The surfaces are computationally modelled and results predict that these rough surfaces create local gradients of E^2, significantly increasing the magnitude of the DEP force throughout the microfluidic channel when compared to conventional 2D-flat electrode-based DEP topologies. Moreover, since the voltage requirement for the stimulation signal is very low, the approach presented in this work is appropriate for integration with microelectronic stimulation systems, which goes in line with the concept of portability in LOC devices.

391669 Simulation and Modeling of Insulator Based Dielectrophoresis

Abstract:

Microfluidics is an emerging field of research with broad range of applications in pharmaceutical bio-analytical and biomedical applications. Insulator based dielectrophoresis is particularly suitable for manipulating biological cells because of the low applied voltages and no direct contact of the cells with the electrodes. Electrokinetic and dielectrophoretic forces have been applied in these devices using various direct and alternate current sources to successfully trap, concentrate and manipulate particles and cells with a significant control. Simulation of the electric field and understanding the forces is critical in further developing and tailoring the designs for a suitable application. Most of the existing models describe forces acting on a single particle and ignoring the effect of particles on the electric field distribution. Present study highlights the discrepancy observed in magnitude of forces between modeling and experiments. The effect of the presence of trapped particles on the electric field distribution is studied with the help of 3D models and simulation. Experimental results on the effect of particle concentration and size on the ability of the microfluidic system to trap will be presented to support simulation studies.

Electrical impedance spectroscopy has been shown to provide a real time evaluation of cell-cell interaction and cell motion [1]. Recently, the application of this technology has been visible in the areas of vascular research, including endothelium hemostasis [2]. However, the majority of published work has been primarily with static cell culture models or macro-scale models that do not properly represent the physiological sizes and structures of human blood vessels [3].

Within this work, we present a microfluidic platform that replicates the hemodynamic loading, as well as, the length scale and structure of a blood vessel with the ability to electrically characterize the endothelium morphology and permeability. Human umbilical vein endothelial cells (HUVECs) are used as an initial endothelium model and are cultured on top of micro-patterned gold electrodes. The microfluidic chip consists of two PDMS microfluidic channels, each with an array of independently addressable gold sensing electrodes of various diameter (50 μm, 100 μm, and 200 μm), and a common large counter electrode. By using a counter electrode that is at least 100 times larger in surface area than the sensing electrode, the impedance measurement is mainly due to the sensing electrode while the counter electrode impedance is negligible. The two microfluidic channels are adjacent and identical, where one acts as a control (no cells) and the other contains cultured cells.

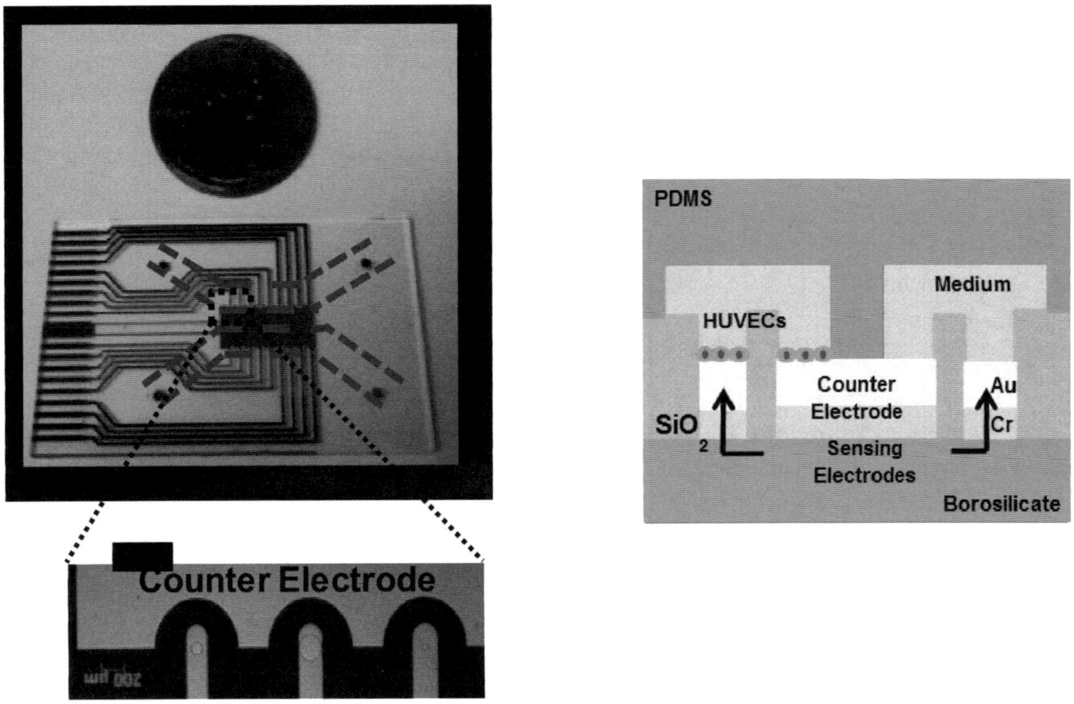

(a) (b)

Figure 1: (A) A top view of the microfluidic chip showing the outline (highlighted in blue) of the microfluidic channels and micro-electrode electrode array. (B) A side view of the chip, identifying the metallic and insulating layers, as well as, the control channel (no cells) and channel designated the cultured cellular monolayer.

The chip is connected to a custom made circuit board that allows automated data acquisition for the different electrodes. Impedance measurements are acquired at a large frequency range (40 Hz- 110 MHz). A syringe or peristaltic pump is also connected to the chip and subjects the cultured human umbilical vein endothelial cells to different shear stress (0-50 dyne/cm^2). Impedance spectra are collected for the following cases: media alone (static and with flow), attachment factor and media (static and with flow), and media with cells (static and with flow). This is done with the purpose to break down the data, determining the effects of medial fluid velocity, attachment factor from that of the cell behavior. Different electrical parameters of HUVEC monolayer were extracted by fitting the impedance spectrum data to an equivalent circuit model using ZView analysis software.

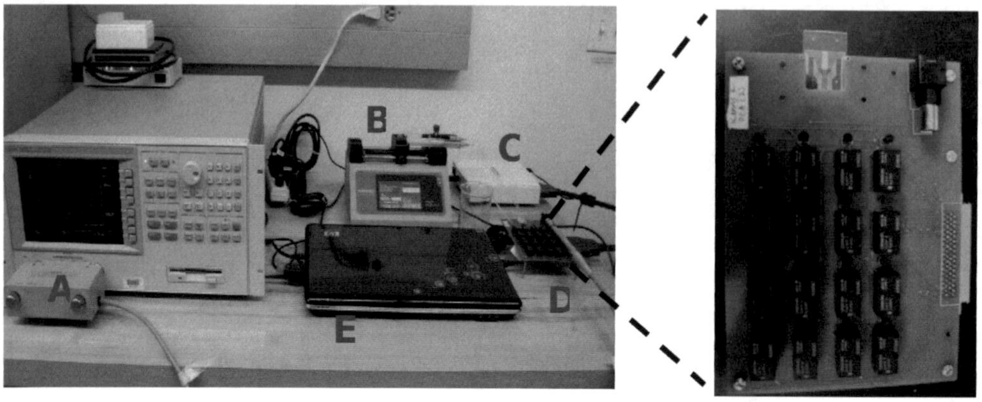

0

Figure 2: The automatic electrode-selecting data acquisition system consists of five main components: (A) An impedance analyzer (Agilent HP4294a), (B) A syringe pump (Harvard Apparatus, Model Pico Plus), (C) A computer interfaced to a multifunction DAQ (National Instruments USB-6251) and data acquisition software (LabVIEW), (D) A custom circuit board for automated channel selection, and (E) Controlled by a custom-made algorithm.

Cell culture media (2700 mS/m) was injected into the microfluidic chip and allowed to equilibrate. Impedance spectra was acquired for 3 sets of electrodes (50, 100, 200 μm) and averaged. Figure 3 shows that as the surface area of the electrode increases, the impedance measurement decreases. As a result, you are more likely to see greater changes in impedance ~~as a result of~~due to cell presence in the bigger electrodes than the smaller electrodes. Larger electrodes will represent the overall cell population. Smaller electrodes will describe a more localized behavior of small population of cells.

Impedance Spectra of cell culture medium was collected at different shear stress (0-50dyne/cm^2) for electrode sizes. As flowrate —(shear stress) increases, impedance increases slightly at the lower frequencies (~10^2-10^5 Hz). At higher frequencies (10^5-10^7 Hz), flow rate has no effect on the impedance. This trend is more noticeable in the smaller electrodes as can be seen from Figure

Impedance measurements were collected for attachment factor and culture media and a static culture of HUVECs within the microfludic chip. Impedance increases just slightly with the addition of attachment factor. However, quantifiable impedance is observed with the presence of cells especially in the mid-range frequencies.

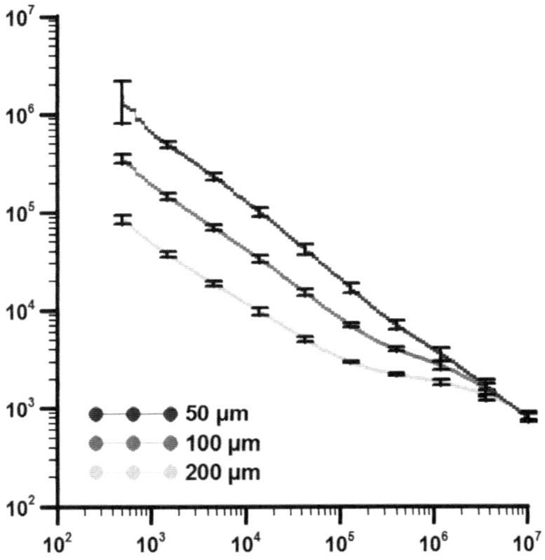

Figure 3: The average impedance spectra of cell culture medium of various electrode diameters.

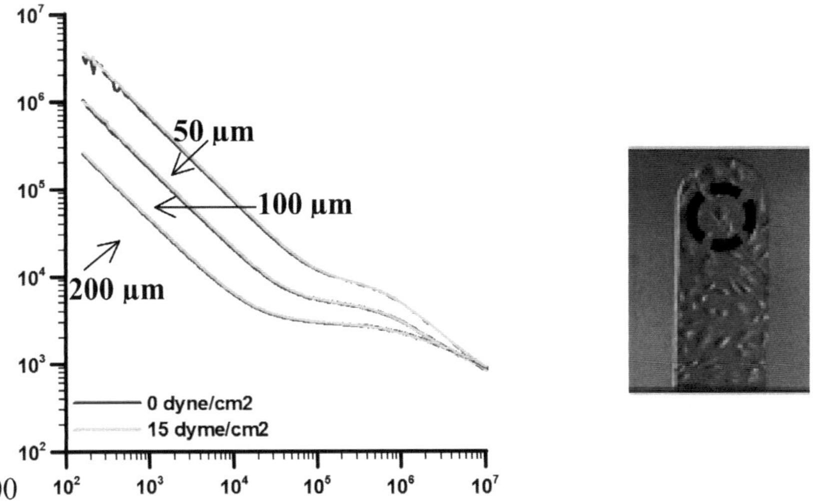

000

Figure 6: (Left) A plot indicating the effects of media flowrate on the impedance spectra of the electrodes. (Right) Image of cultured cells on top of a gold electrode.

The microfluidic chip presented in this work introduces an automated system for the characterization of cellular monolayers under fluid shear. We plan to continue the use of the system for determine any changes in permeability of the cellular layer due to varying shear stress. It can be used as a real-time measurement technique that is less expensive and less complex compared to traditional optical observation methods. The device is suitable for various applications other than cardiovascular disease. For example, it can have a significant impact on characterizing cancer metastasis under mechanical flow, wound healing mechanics, and the testing of new pharmaceutical agents' adsorption and effects under physiologically relevant fluid dynamic conditions.

5. REFERENCES

1. Giaever, I., Keese, C.R. "Micromotion of mammalian cells measured electrically" *Proceedings of the National Academy of Sciences of the United States of America* 88(17) (1991):7896-7900.
2. DePaola, N., Phelps, J.E., Florez, L., Keese, C.R., Minnear, F.L., Giaever, I., Vincent, P. "Electrical impedance of cultured endothelium under fluid flow" *Annals of Biomedical Engineering*, 29(8) (2001):648-656.
3. Young, E.W., Simmons, C.A. "Macro- and microscale fluid flow systems for endothelial cell biology" *Lab on a Chip*, 10(2) (2010):143-160.

389019 Study of Cell Viability after Manipulation with Insulator-Based Dielectrophoresis

Abstract:

Food and water safety, environmental monitoring and clinical screening are examples of fields where microfluidic devices or lab-on-a-chip systems can make a significant impact, since in these areas obtaining a rapid response is critical, particularly in analyses involving biological cells. Microfluidics has revolutionized the manner in which many assessments/processes are carried out. Biotechnology is one of the fields that has been most benefited with the marriage of microfluidics and analytical sciences. Many biotechnological applications require fast response methods that are able to handle/analyze high value biological products, such as cells, in a gentle manner, without leading to cell damage.

Insulator based dielectrophoresis (iDEP) is a leading microfluidics technique employed for the concentration, sorting and separation of samples containing cells. In this technique cells are manipulated inside microchannels by utilizing non-uniform electric fields. Briefly, microchannels for iDEP contain an array of 3-dimensional insulating structures, electric field are across the microchannel by employing electrodes located at the channel inlet and outlet. The presence of the insulating structures creates a non-uniform electric field enabling dielectrophoresis. Many successful studies have been published on the analysis of cell using iDEP. However, during iDEP manipulation cells are exposed to high electric field gradients and Joule heating, which affects cell viability. The present study will quantify the decrease of cell viability after exposure to iDEP manipulation. Both E.coli and Yeast were tested in identical iDEP microdevices only varying in post geometries. Six different post geometries were analyzed, three cylindrical and three diamond, at three different time durations.

Low Cost Fabrication of Microchannels for Lab-on-a-Chip Applications

Monsur Islam, Rucha Natu and Rodrigo Martinez-Duarte*

Mechanical Engineering Department, Clemson University, SC, USA. Contact: rodrigm@clemson.edu

CLEMSON UNIVERSITY

M²L MULTISCALE MANUFACTURING LABORATORY

Introduction

Here we present detailed characterization of a low cost method for fabricating microchannels for lab-on-a-chip devices using a desktop digital cutting plotter and double sided pressure sensitive adhesives (PSA). Our aim is at offering the community an inexpensive complement to soft lithography. To this end, the limits of the proposed process are reported here in terms of dimensions and shapes that could be fabricated.

PSAs are inexpensive and commercially available from different vendors in a variety of presentations. A professional-grade cutting plotter is under $1,500. This process does not require master molds, degassing of a polymer or heating sources.

Channel Geometries

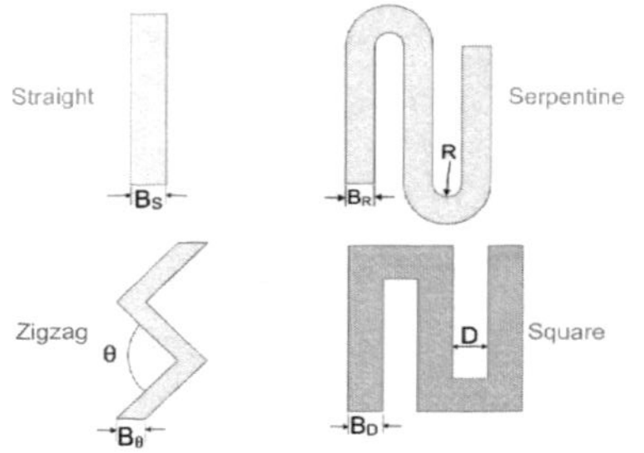

Fabrication Process

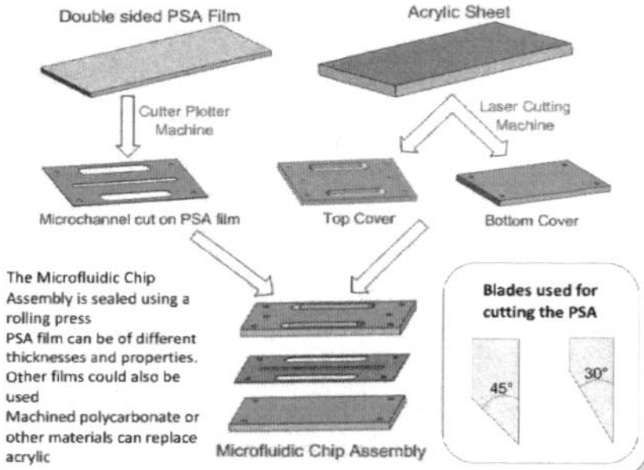

- The Microfluidic Chip Assembly is sealed using a rolling press
- PSA film can be of different thicknesses and properties. Other films could also be used
- Machined polycarbonate or other materials can replace acrylic

What could YOU do?

For example, Microfluidic Chips for:

- ✓ Electrophoresis and Dielectrophoresis
- ✓ Electroosmosis
- ✓ Gradient generators
- ✓ CD microfluidics

Examples of Devices:

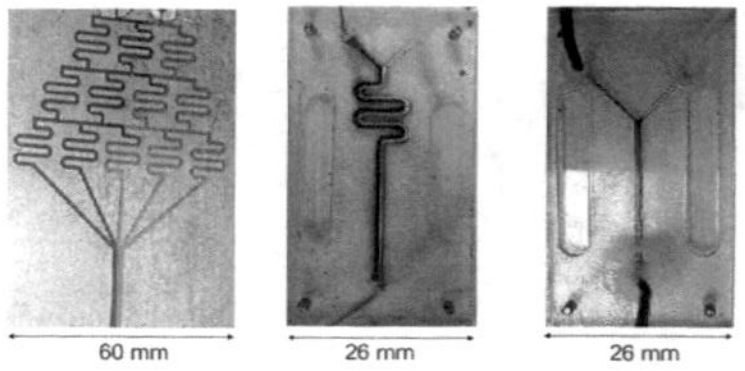

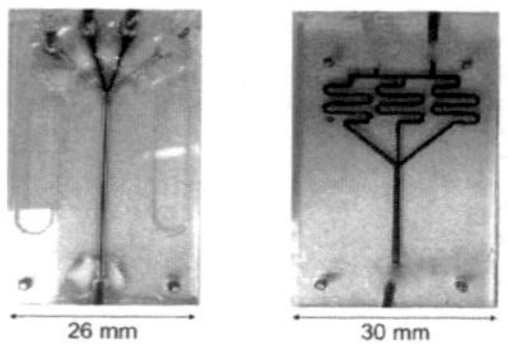

Gradient generators of different designs for mixing purposes

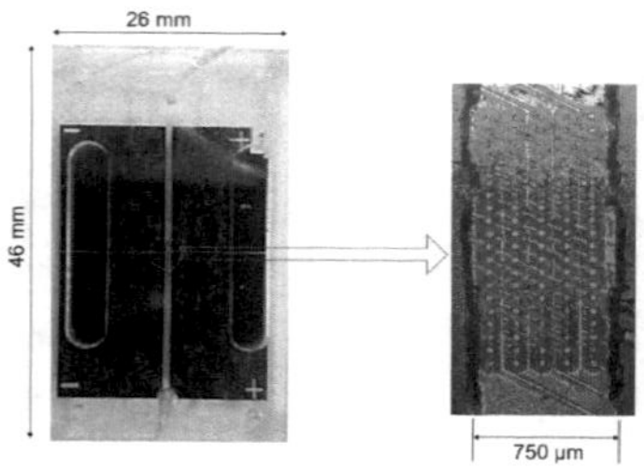

Microfluidic chip for performing dielectrophoresis experiments (note the roughness of the channel around the circular carbon electrodes)

Results and Observations

For PSA thickness = 127µm, Cutting Parameters: Speed = 1cm/s, Force = 19N, Acceleration = 1cm/s²

Straight

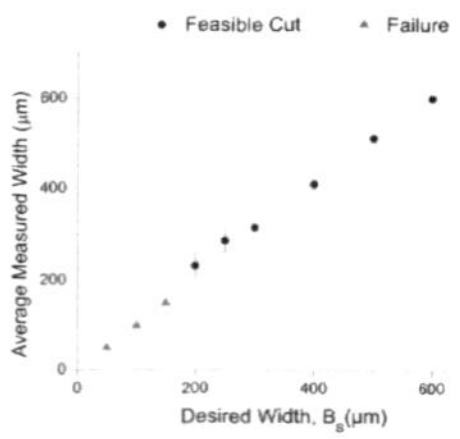

Serpentine

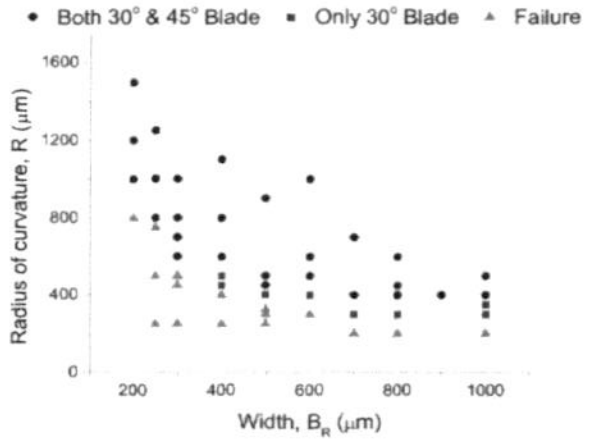

Zigzag

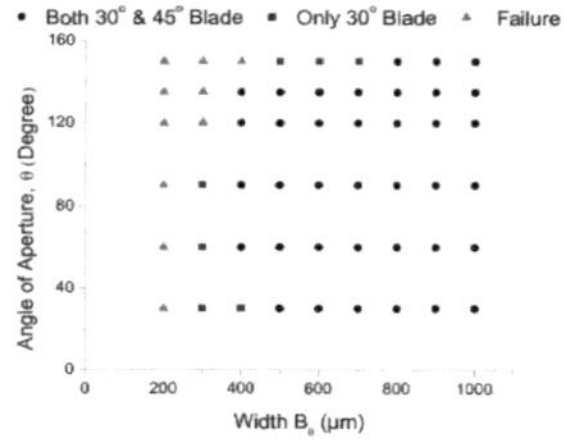

Square

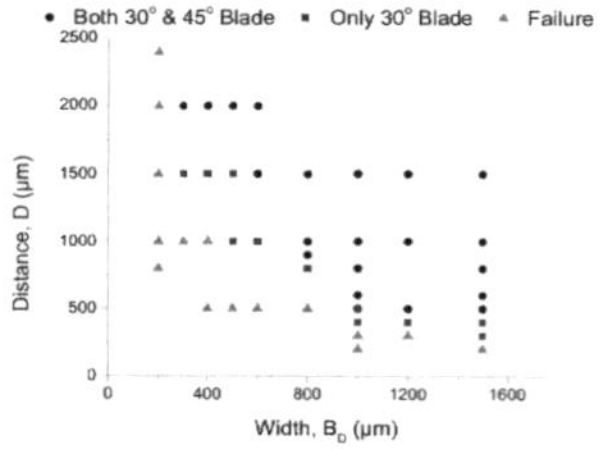

General Observations:

- Channels below 300 µm-wide may cut as a single line
- Designs with channel width < 300 µm were drawn as individual lines instead of a contour
- In zigzag channels, an angle θ > 120° leads to an abrupt channel width dimension
- Quality of the channel wall depends on the cutting parameters
- Flat PSA sheets tend to work better than rolled sheets
- Adhesive thickness > 300 µm may require higher multiple cut passes, with mixed results

Advantages:

- No need for cleanroom
- Simple and inexpensive
- Idea-to-device time of minutes
- Enables the use of very high flow rates (ml/min)
- Wide selection of functionalized adhesives available
- Different materials can be used as the top and bottom cover
- Chip assembly can be dismantled and some parts re-used

Disadvantages:

- Channel walls may be rough for some applications
- Accuracy sharply decreases below 300 µm
- Precision decreases below widths of 600 µm
- Successful cuts can depend on the residual stress in the adhesive films
- Parts of the walls may come off at very high flow rates (>15 ml/min)

Acknowledgements:

The authors thank Alicia Werner, Emily Wakefield and Madison Oldham of the Creative Inquiry program in Clemson University for pictures and discussion regarding gradient generators

Visit: www.multiscalemanufacturing.net

365173 Electrokinetic Particle Sorting By Shape in a Spiral Microchannel

Abstract:

Shape is an intrinsic marker of cell cycle and also a useful indicator of cell state for disease diagnostics. We present in this talk a continuous-flow electrokinetic sorting of spherical and peanut-shaped particles of similar volumes in an asymmetric double-spiral microchannel. It exploits curvature-induced dielectrophoresis (C-iDEP) to focus particles to a tight stream in the first spiral and subsequently displace them to shape-dependent flow paths in the second spiral. We also develop a numerical model to simulate and understand this shape-based particle sorting in spiral microchannels, which predicts the experimental observation qualitatively.

365178 Electrokinetic Instabilities in Ferrofluid Flows

Abstract:

Instabilities in ferrofluids have so far been reported under the application of a magnetic field, which can be utilized to, for example, mix samples at small Reynolds numbers for microfluidic applications. We present in this work our recent experimental observations of instabilities occurring at the interface of electric field-driven ferrofluid and water co-flows. Such electrokinetic instabilities are produced primarily by the mismatch of electric conductivities of the two fluids. A two-dimensional numerical model is developed to compute the electric, flow and concentration fields, which is able to qualitatively simulate the onset and pattern of electrokinetic instabilities.

The impact of electrode geometry in Dielectrophoretic effect for multilayer PPyDEP-based devices

*Victor H. Perez-Gonzalez,[a] Vinh Ho,[b] Sergio O. Navarro-Rodriguez,[a] Lawrence Kulinsky,[b] Marc J. Madou,[a,b,c] and Sergio O. Martinez-Chapa,[a]**

a Department of Electrical and Computer Engineering, Tecnológico de Monterrey, Nuevo León, 64849, México.
b Department of Mechanical and Aerospace Engineering, University of California Irvine, California, 92617, United States.
c Department of Biomedical Engineering, University of California Irvine, California, 92617, United States.

*Correspondence author. Email: smart@itesm.mx

Dielectrophoresis (DEP), the electrokinetic technique capable of manipulating polarizable particles in the presence of a spatially non-uniform electric field, has attracted significant attention from the scientific community over the past two decades. This technique has many important advantages over conventional microparticle manipulation approaches; for instance, DEP has great potential for miniaturization, it is a non-invasive and non-destructive manipulation technique, it only requires small volumes of sample, and it is a selective technique capable of separating, concentrating, filtrating, or characterizing microparticles. Research fields such as as environmental monitoring, food quality assessment, clinical analysis, among many others have already benefited from the attractive features of DEP.

Currently, two main branches exist for the design of DEP-based microdevices, namely electrode-based DEP and insulator-based DEP (iDEP). Electrode-based DEP has largely been associated with 2D electrode designs (interdigitated, polynomial, or castellated). Microfabrication techniques allow the gap between electrodes to be very narrow for these designs, lowering the requirement for large excitation voltages and therefore, decreasing the effect of Joule heating as well. Unfortunately, 2D electrodes only induce DEP in particles that are located in close vicinity to them, leaving most of the particles flowing in the bulk of the solution unaffected. In contrast, iDEP overcomes such limitations through the fabrication of 3D electrical-insulating obstacles that distort the distribution of the electric field generated within the microfluidic channel. The main drawback of iDEP devices is that very large excitation voltages (e.g., 1200V) are required to induce DEP in the particles. As a result of such excitation voltages, Joule heating is generally associated with iDEP devices. Recently, CarbonDEP and PPyDEP have emerged as promising 3D electrode-based dielectrophoretic alternatives. These techniques bring together the benefits (and overcome the limitations) of conventional electrode-based DEP and iDEP.

This work studies the impact of electrode geometry in the dielectrophoretic effect observed in multilayer PPyDEP-based devices. To accomplish this, a novel multilayer electrode design is presented, computationally analyzed, and compared to results readily available in the literature for PPyDEP devices. Characterization of the device allows predicting the location of the trapping zones for both, positive and negative DEP. Moreover, the dielectrophoretic effect over polystyrene beads of different sizes along several regions of the device are also presented. The predictions obtained from this study demonstrate that the proposed multilayer electrode design provides significant advantages over its preceding version. For instance, the average magnitude of the dielectrophoretic force inside the microfluidic channel is increased by at least an order of magnitude, having a positive impact in the volume of sample affected by the DEP force.

M.D. - ERL
www.mderl.org

Simultaneous Detection and Quantification of Water- and Fat-Soluble Vitamins with Liquid Chromatography and Tandem Ion Trap-Mass Spectrometry

M. Khaksari[1], L. Mazzoleni[2], C. Ruan[3], P. Song[4], N. Hershey[4], R. Kennedy[4], M. Burns[5], A. Minerick[1]

[1]Department of Chemical Engineering, [2]Department of Chemistry, Michigan Technological University, Houghton, MI
[3]Metabolomics Core, BRCF, [4]Department of Chemistry, [5]Department of Chemical Engineering, University of Michigan, Ann Arbor, MI

MichiganTech

Background and Motivation

- Obesity is a form of malnutrition.
  → Need food rich in nutrients without empty calories.

Obesity in USA in 2013 [1]

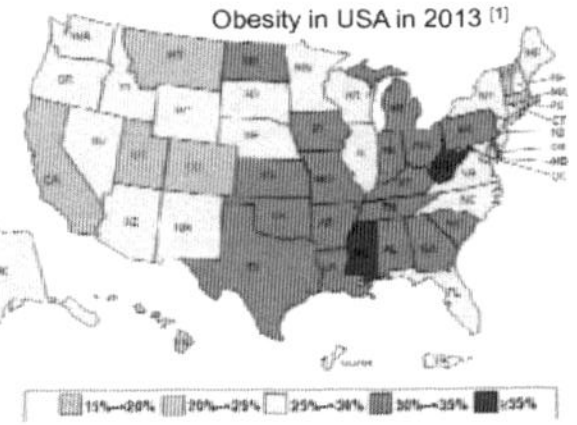

- Big Picture: rapid and sensitive vitamin detection can indicate nutritional health and provide feedback.
  → This work describes development and optimization of a combined HPLC/MS protocol.

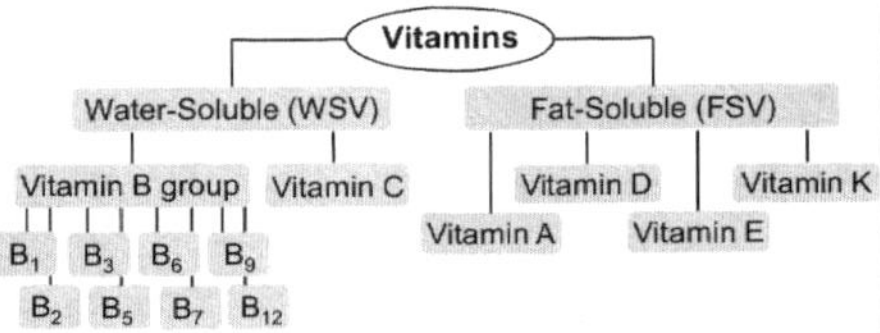

- Stable isotopes of vitamins used as internal standards (IS) to reduce the quantification errors.

Method Conditions

High Pressure Liquid Chromatography (HPLC) Conditions:

- Thermo Accela HPLC autosampler and pump
- C_{18}, silica-based, 2.1 x 150 mm column
- Injection volume: 25 µL
- At 0.2 mL / min
- Mobile phases:
  - mpA: 0.1% formic acid in water
  - mpB: 0.1% formic acid in ACN
  - mpC: 5 mM ammonium formate in MeOH

Mass Spectrometry (MS) Conditions:

- Thermo LCQ Fleet 3D linear ion trap with Electrospray Ionization (ESI) probe
- Sheath gas flow rate: WSV (10) / FSV (20)
- Spray voltage: 4 kV
- Capillary temperature: 275°C

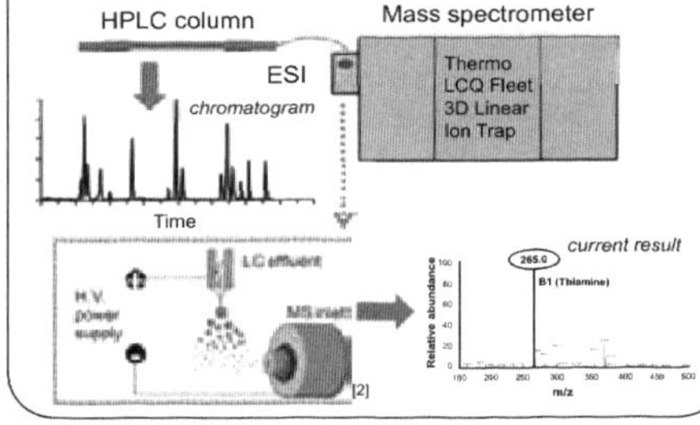

Protocol Results

HPLC chromatograms of all WSV and FSV

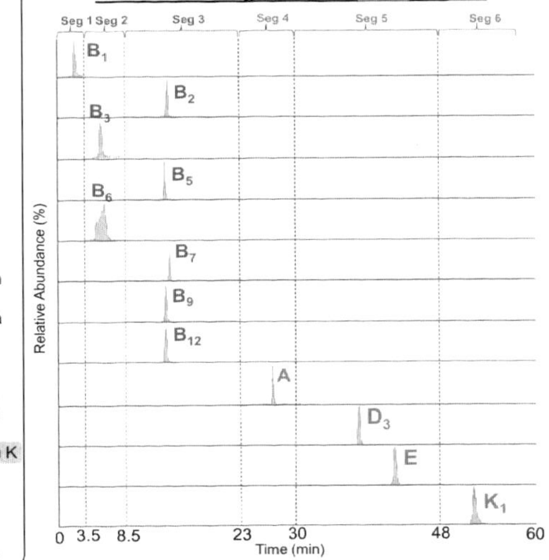

- All vitamins were detected simultaneously using 6 segments within 53 min.
- Peak areas are proportional to vitamin amounts.

Optimized gradient elusion of mobile phases

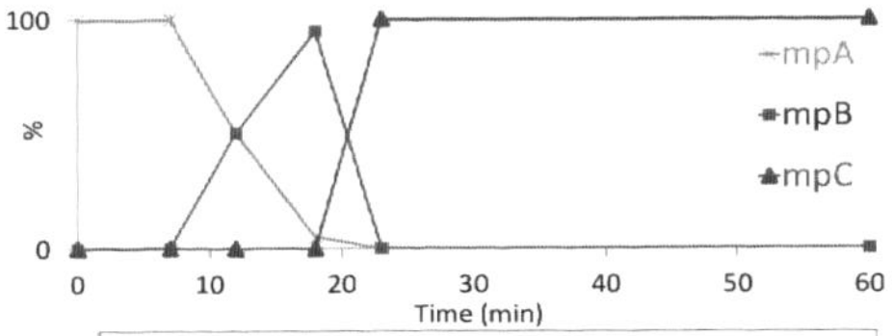

- Three mobile phases eluted all vitamins; mpA and mpB eluted WSV and mpC eluted FSV.

MS results of WSV and FSV and their IS

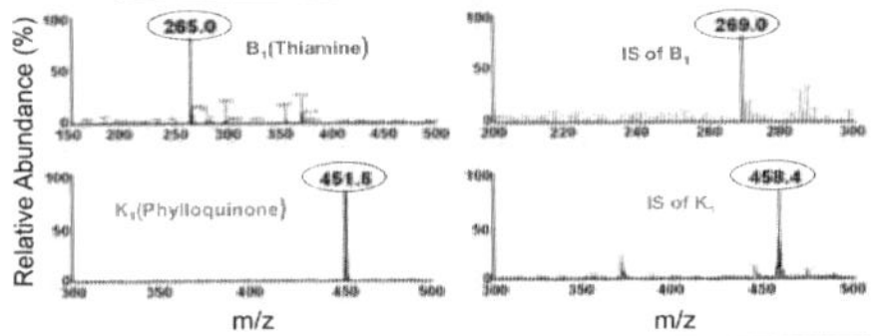

- Protonated ions of vitamins were used for detections.
- Fragment ions were used for quantifications.

Study of adduct formation for ionization of FSV

- **Additives tested:** silver perchlorate, cesium acetate, ammonium formate, ammonium acetate, formic acid.

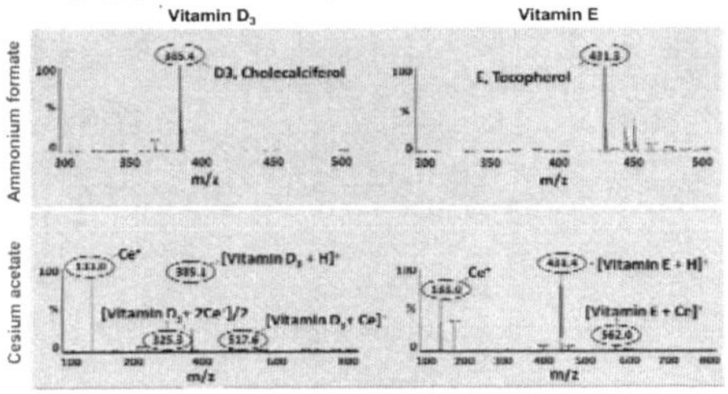

- 5 mM ammonium formate was added to mpC.
- Adducts by cesium acetate, silver perchlorate, ammonium acetate and formic acid were not intense enough for reliable detection.

Calibration curves for WSV and FSV used to find concentrations in blood serum:

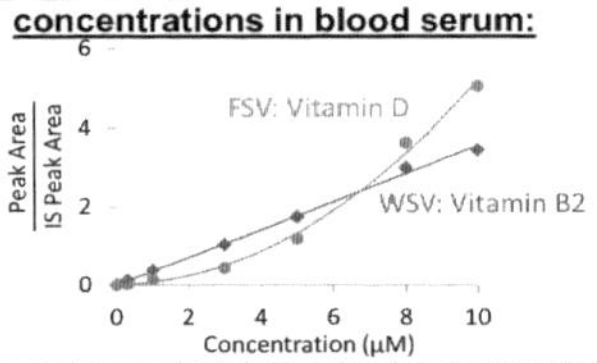

Vitamins	Calibration curves	R^2	LOD (ng)	LOQ (ng)	Linear range (µmol/L)	Intra-day (% RSD)	Inter-day (% RSD)
B_1, Thiamine	y = 1.151x + 0.001	0.9957	0.043	0.14	0.02-50	4.6	7.9
B_2, Riboflavin	y = 0.3556x + 0.001	0.9970	0.055	0.18	0.02-100	2.5	2.3
B_3, Nicotinamide	y = 0.06850x + 0.003	0.9950	0.60	2.0	0.7-100	2.1	4.6
B_5, Pantothenic acid	y = 0.2919x + 0.001	0.9930	0.60	2.0	0.4-200	8.8	7.2
B_6, Pyridoxine	y = 0.3557x + 0.006	0.9998	0.082	0.27	0.05-200	2.3	8.5
B_7, Biotin	y = 0.2611x + 0.006	0.9989	0.30	0.99	0.2-10	2.5	12.7
B_9, Folic acid	y = 5.236x + 0.344	0.9910	0.61	2.0	0.2-100	9.7	6.1
B_{12}, Cyanocobalamin	y = 4.415x + 0.022	0.9912	1.2	4.0	0.1-10	9.0	8.6
A, Retinol	$y = 0.0065x^2 - 0.0042x + 0.002$	0.9995	9.6	15	-	4.0	5.6
D_3, Choleocalciferol	$y = 0.050x^2 + 0.020x + 0.002$	0.9949	1.1	2.77	-	2.9	5.6
E, α-Tocopherol	$y = 0.69x^2 + 0.51x + 0.02$	0.9982	0.13	0.40	-	2.8	7.2
K_1, Phylloquinone	$y = 0.012x^2 + 0.14x + 0.001$	0.9836	0.23	0.77	-	1.0	4.6

Method Validated with Blood Serum

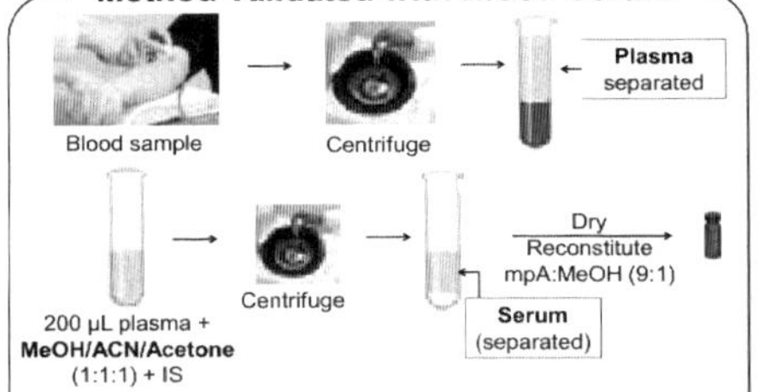

WSV and FSV chromatograms detected in human blood serum:

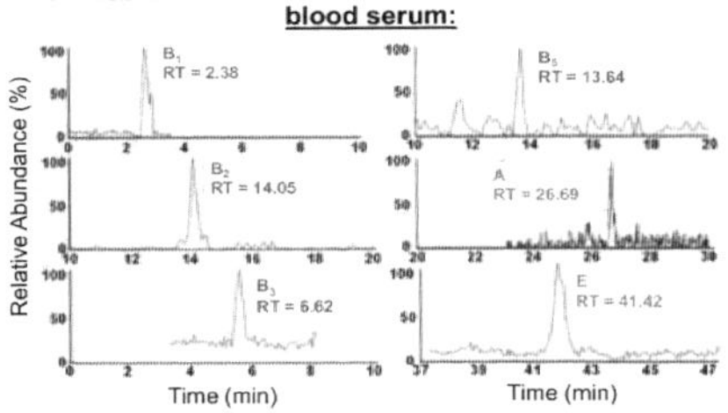

- Vitamins concentrations in this serum sample calculated from WSV and FSV calibration curves:
  - B_1 = 0.016 µM, normal range [3]: 0.009-0.044 µM
  - B_2 = 0.12 µM, normal range [3]: 0.003-0.050 µM
  - B_3 = 1.46 µM, normal range not reported.
  - B_5 = 0.83 µM, normal range [3]: 0.016-3.76 µM
  - A = 0.92 µM, normal range [3]: 0.98-3.3 µM
  - E = 4.61 µM, normal range [3]: 8.82-42.7 µM

Conclusions

- Combined HPLC/MS method developed for determination and quantification of WSV and FSV.
  - Prior published work utilized a subset of vitamins [4,5].
  - Provides substantial time savings for nutritional assessments.
  - All WSV and FSV eluted under 53 min with ternary mobile phases.
  - LOD and LOD comparable to literature.
  - Intra- and inter-day precisions were good from 1.0-12.7 %RSD.
- Method successfully validated with human blood serum.

References

1. Overweight and Obesity [Online], in: Obesity Prevalence Maps. Center for Disease Control and Prevention (CDC), 2013.
2. Song, P., Mass spectrometry based techniques for in vivo neurochemical monitoring. PhD Dissertation, 2012. 111: p. 12.
3. http://www.mayomedicallaboratories.com/interpretive guide, access in June, 6, 2014
4. Chen, Z., B. Chen, and S. Yao, High-performance liquid chromatography/electrospray ionization-mass spectrometry for simultaneous determination of taurine and 10 water-soluble vitamins in multivitamin tablets. Analytica Chimica Acta, 2006. **569**(1–2): p. 169-175.
5. Priego Capote, F., et al., Identification and determination of fat-soluble vitamins and metabolites in human serum by liquid chromatography/triple quadrupole mass spectrometry with multiple reaction monitoring. Rapid Commun Mass Spectrom, 2007. **21**(11): p. 1745-54.

Acknowledgements

Dr. D. Burke at U. of Michigan Human Genetics Department for valuable discussions.

395930 Improving the Design of Insulator-Based Dielectrophoretic Devices

Abstract:

Dielectrophoresis (DEP) is an electric field driven technique that originates from polarization effects when particles are exposed to non-uniform electric fields. Insulator-based dielectrophoresis (iDEP) employs insulating structures embedded in a microchannel to produce electric field gradients necessary to generate dielectrophoretic forces on particles. This contribution presents a detailed analysis of particle trapping capacity in microchannels with arrays of insulating structures. The effect of the shape, dimensions and spacing of the insulating structures on the dielectrophoretic trapping were explored. This study includes experimental and mathematical modeling work. COMSOL Multiphysics® was used to determine the distribution of the electric field and electric field gradient within the array of insulating posts in order to determine the optimum set of parameters that maximize the particle trapping capacity of a channel. The phenomena that contribute to particle movement are introduced with an analysis of particle motion with respect to a local electric. Experimentation was performed by applying DC electric potentials across an iDEP microchannel designed with the optimum set of parameters and comparing its performance with commonly used microchannels. All experiments were carried out with devices made from PDMS. The experimental and modeling results clearly depict the dependence of the particle trapping capacity with respect to the considered design parameters. The findings from this study provide a systematic approach towards the design of highly efficient iDEP devices.

395983 Dielectrophoretic Behavior of Polystyrene Particles Under Direct Current and Low Frequency Electric Fields

Abstract:

Dielectrophoresis (DEP) is the movement of particles due to polarization effect when particles are exposed to non-uniform electric field. In insulator based dielectrophoresis (iDEP), non-uniform electric fields are generating by employing arrays of insulating structures. The Clausius-Mossotti (CM) factor is the parameter that accounts for particle polarization and it can be positive or negative, the CM factor dictates the direction of the dielectrophoretic force, which can be positive or negative. Positive DEP (pDEP) is when particles are attracted to the regions of higher electric field gradient, and negative DEP (nDEP) is particle repulsion from those regions. When the Clausius-Mossotti (CM) factor for polystyrene particles is plotted as a function of electric field frequency, high frequencies correspond to negative CM factors and negative dielectrophoresis (nDEP) while low and mid-range frequencies correspond to positive CM factors and positive dielectrophoresis (pDEP). However, iDEP experiments with direct current (DC) electric fields generally result in nDEP even when the particles are supposed to have positive CM factors. This work is focused on characterization the behavior of polystyrene particles of different sizes, including sub-micron level, under DC and low frequency alternating current (AC) electric fields. Modeling with COMSOL Multiphysics was used to theoretically predict the distribution of the electric field in microchannel with insulating posts. Experiments were performed in devices made from polydimethylsiloxane (PDMS) through soft lithography techniques. The results will provide a deeper understanding of dielectrophoretic particle behavior in the conductivity regime.

397149 Rapid Electrokinetic Patterning: Manipulating Particles with Laser and Electric Field

Abstract:

Rapid Electrokinetic Patterning (REP) combines optical and electrokinetic effects to create individually addressable traps for micro- and nanoparticles. By projecting reconfigurable light patterns, particles can be concentrated, manipulated and sorted in a programmable manner. Previously, this technique has been used for trapping of various dielectric particles. In this poster, we show that the metal nanoparticles such as gold and silver nanospheres (40-200 nm), silver nanowires and multi-wall carbon nanotubes can also be reversibly trapped using this method. We demonstrate application of REP in sensitive chemical detection with Surface Enhanced Raman Spectroscopy (SERS). SERS requires metal nanoparticles to be concentrated in a controlled manner which is achieved by REP induced aggregation.

399322 Electrokinetic Separation of Cells Using Ionic Liquids (ILs) Catalysts

Abstract:

Characterization and analysis of rare cells demonstrate critical cues for early diagnosis of chronic diseases. Electrokinetic cell separation methods have been previously established to have greater efficiency when compared to traditional flow cytometry methods. It has been shown by many researchers that buffer solutions in which cells are suspended in have enormous effects on producing required dielcctrophoretic forces to separate and manipulate cells. Most commonly used suspension buffers used to electrokinetically manipulate cells are deionized water and cell media. However, these solutions exhibit high level of intrinsic noise, which greatly masks the electrokinetic signals from cells under study. Ionic liquids show promise towards the creation of conductive fluids with required electrical properties. The goal of this project is to design and test tailored ionic liquid compositions for enhancing dielectrophoretic forces on cells while creating an environment for preserving their integrity. We analyzed three uniquely synthesized methylimidazolium based ionic liquids with low half maximal inhibitory concentration was used as suspension medium for cell separation. These dicationic ionic liquids possess slight electrical and structural differences with high thermal stability. The three ionic liquids were tested for cytotoxicity and their ability to enhance signals from cells. We performed tests using organic particles (HeLa cells) and inorganic particles (synthetic beads). Separation yield, system efficiency, purity and recovery of isolated cells were calculated including cell viability. The effects of electrical neutrality, charged particle effects, free charge screening due to ionic liquids and their contribution towards enhanced electrokinetic signals from cells were studied. We were able to achieve improved system resolution (upto 95%) with no harmful effects on cells.

399333 Identification and Characterization of Rare Cells through Electrokinetic Cell Oscillations

Abstract:

The goal of this project is to demonstrate an electrokinetic tool for label-free detection, isolation, characterization and analysis of rare cells based on their biomolecular profiles.

Different normal cells and rare cells undergo different cellular processes based on their unique biomolecular profile. Consequently they modulate the overall electrical properties of the cells and their ability to polarize under external electric fields. For normal cells, the dielectric properties are well characterized based on their internal and surface biomarkers. Hence they can be used to electrically control and isolate them in biological samples. However, apriori knowledge of rare cells such as cancer stem cells, circulating tumor cells, progenitor cells, etc, are not well-known. Recent scientific advances have demonstrated the proof of concept in detecting and isolating rare cell population with the little known knowledge of their cell surface proteins and markers. However, the ability to distinguish between endothelial stem cells and endothelial progenitor cells are very marginal using present day methods.

Here we present a successful label-free lab-on-chip tool which could eventually be used as an add-on module to existing methods to identify and characterize based on their biomolecular profile. The platform utilizes size-matched microelectrodes to apply gradient electric fields to the system and microfluidic channels for sample fluid flow. Cells under study are suspended in conductive medium to facilitate differential polarization inside and outside of the cell. As a proof of concept, we demonstrate the applicability of this platform tool for cancer stem cells which have demonstrated positive activity of aldehyde dehydrogenase (ALDH) isozymes. In this study, we used a simulated ratiometric mixture of ALDH+ and ALDH- cell lines. The cells were derived from synthetically cultured lung cancer cell lines enriched for cancer stem cells. ALDH+ cells have a unique biomolecular profile that affects its polarizability which results in resonance behavior at harmonic frequencies.

Upon characterization through electrokinetic oscillatory tool, we were able to perform cell counting and successfully distinguish cancer cells that were ALDH+ and ALDH- cells. Image processing algorithms were used to characterize the results in less than 15 minutes post introduction of input sample. Statistical modeling were performed to estimate the probability of success rate. In conclusion, the developed tool was able to identify and isolate target cells with significance of $p<0.05$, leading to efficiency of separation upto 90%. In conclusion, we present a method to identify rare cells based on their unique oscillatory behavior as a function of their biomolecular profile.

399344 Reproducibility Using the Amersham™ WB System

Abstract:

The Amersham WB system is an integrated system for SDS PAGE and Western blotting based on fluorescent detection. The system is composed of proven technology and reagents. The workflow is standardized in every step from sample application to evaluation of results to minimize user to user variation. The separation, transfer, probing and detection has been carefully optimized to obtain high signal to noise with high consistency. The system was designed to produce reliable, quantitative and reproducible results every time. Results are presented for SDS PAGE and Western blotting showing high reproducibility within or between gels/membranes for different users, both experienced and first time users.

Exploiting Absolute Negative Mobility with Dielectrophoresis for Mitochondrial Sample Preparation

Jinghui Luo,[1] Alexandra Ros[1]

[1]Department of Chemistry and Biochemistry, Arizona State University, Tempe, AZ 85287.

The study of mitochondria is important due to the organelle's contribution to many cellular functions, including metabolism, cellular signaling and apoptosis. The atypically sized giant mitochondria[1] were found in the models of aging and some diseases. Thus it would be advantageous for related studies if mitochondrial subpopulations could be prepared in accordance with their particular sizes. Conventional methods such as centrifugation and electrophoresis can separate mitochondria from cell cytoplasm and further purify them from other organelles. However, the separation of mitochondrial subpopulations has rarely been studied. Here, we present the numerical simulation with the design of a microfluidic device to elucidate absolute negative mobility (ANM) utilizing insulator-based dielectrophoresis (iDEP) for the separation of mitochondrial subpopulations.

ANM has recently been proposed as a separation mechanism for micron-sized particles[2]. It is the motion of a particle into the direction opposite to a net applied force[3], which can be realized when Brownian particles migrate in a non-linear structure under non-equilibrium conditions. For instance, when particles migrate with flow (y direction in Figure 1a) in an ANM structure, they could be trapped by adjacent posts which form a gap smaller than their diameter. On the other hand, probability exists that they can also avoid the trap due to their diffusion in the direction perpendicular to the flow. This probability for a Brownian particle - being trapped or avoiding a trap - relates to the diffusion property of the particle, i.e. its size. Therefore, under non-equilibrium conditions particles within a certain size range may move towards the direction of a net force while other particles may move in the opposite direction and consequently, separation based on ANM may be realized. The design of the ANM structure has been studied previously both in theory with the variation of geometric post shapes and in experiment with micron-sized colloidal particles in a microfluidic device[4]. However, this approach is not transferrable to evoke ANM for the separation of mitochondrial subpopulations with typical sub-micron sizes (0.1 – 1 μm)[1]. Thus, we additionally exploit the dielectrophoretic properties of this important organelle in tailored microdevices in order to generate ANM. DEP is a migration phenomenon based on the polarization property of a particle in an inhomogeneous electric field. This property has been used to separate, concentrate or distinguish many biological species, including cells, tissue, proteins, bacteria and DNA[5]. Our previous studies[6] revealed that isolated Fischer 344 rat semimembranosus muscle mitochondria and C57BL/6 mouse hepatic mitochondria exhibited negative DEP (nDEP) under both DC (0 – 3000 V/cm) and AC (0 – 2000 V/cm, 0 – 50 kHz) conditions in low conductivity buffer in physiological pH. In addition, the potential thresholds for trapping the mitochondrial samples were weakly depended on the applied frequencies. Based on that, a structure modified to suit the small size exploiting the nDEP property of mitochondria was designed (Figure 1b).

First, the migration properties of particles exhibiting nDEP were studied in numerical simulation using *COMSOL Multiphysics 4.4*. Properties of micron and sub-micron polystyrene beads were used to mimic the mitochondrial subpopulations. Non-equilibrium conditions were realized by an alternating electric field, which was tunable by combining the static study result of the *Electric Currents* module with a rectangle and an analytic function. This was further coupled with the time-dependent study using the *Creeping Flow* module, resulting in an alternating flow profile, which was further integrated with DEP and Brownian forces in the time-dependent study of the *Particle Tracing for Fluid Flow* module. The resultant particle trajectory allowed for the investigation of the conditions such as the applied potential and the dimensions of the posts and the gaps needed for realizing ANM for the separation of micron- and sub-micron sized particles and thus accessing a regime suitable for mitochondrial subpopulations. We found that with certain conditions, ANM could be observed with particles of 5-μm diameter (Figure 1d) while 1-μm particles do not show ANM. This novel approach to induce differential ANM migration behavior for particles of varying sizes allows retaining simple photolithographic device fabrication strategies for experimental ANM applications with organelles, and similarly preserves the size selectivity for sub-micron bio-species.

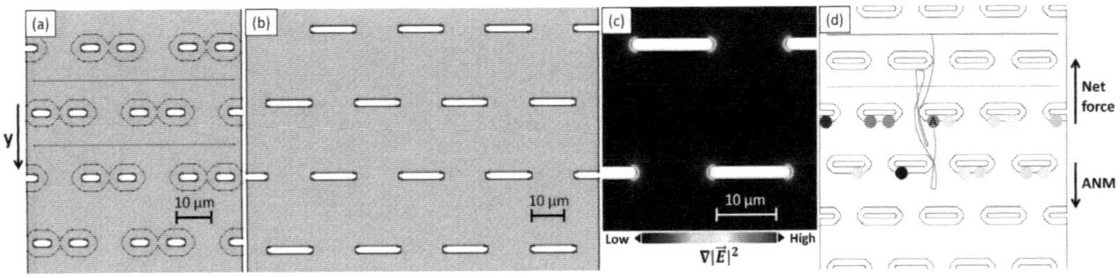

Figure 1: Schematics of the ANM devices. Electric potential was applied along the y-axis. The posts and the channel walls are outlined in blue, while the black lines in (a) indicate the virtual walls which are used to exclude particles from the post regions with a distance equal to or smaller than their radius. (a) The two black horizontal lines indicate the positions where particles are released in the simulation. Because the small gap is 5 μm wide, this exemplary structure is suitable for the ANM study of particles with diameter no smaller than 5 μm. (b) Alternative layout to couple a DEP component. This structure is suitable for the ANM study of mitochondria of which the typical size range is from 0.1 μm to up to 1 μm[1]. (c) Zoom-in of (b) showing the electric field gradient distribution around the posts. If there is flow in the y direction, a particle exhibiting nDEP will be pushed towards and trapped in the center of a post similarly to a geometrical trap as shown in (a). (d) ANM behavior simulated with particles of 5 μm diameter in two consecutive time periods of the applied frequency. Particles started at the two horizontal lines and migrated with flow in the y direction. The trajectory of particle A was labeled in the figure demonstrating ANM.

References

1. M. Navratil, A. Terman, and E. A. Arriaga, Exp. Cell Res. **314,** 164 (2008).
2. A. Ros, R. Eichhorn, J. Regtmeier, T. Duong, P. Reimann, and D. Anselmetti, Nature **436,** 928 (2005).
3. R. Eichhorn and P. Reimann, Acta Physica Polonica B **37,** 1491 (2006).
4. J. Regtmeier, S. Grauwin, R. Eichhorn, P. Reimann, D. Anselmetti, and A. Ros, Journal of Separation Science **30,** 1461 (2007).
5. T. Z. Jubery, S. K. Srivastava, and P. Dutta, Electrophoresis **35,** 691 (2014).
6. J. Luo, B. G. Abdallah, G. G. Wolken, E. A. Arriaga, and A. Ros, Biomicrofluidics **8,** 021801 (2014).

399320 Low Cost Microwave Plasma Generation for the Irreversible Sealing of PDMS Microfluidic Devices-

Abstract:

Sealing of Poly-dimethylsiloxane (PDMS) microdevices onto glass or PDMS substrate is an important step to achieve hydrophillicity and form closed microfluidic channels. PDMS is very hydrophobic and to enhance hydrophilic behavior, PDMS substrates can be treated using a UVO or plasma cleaner. Commercial plasma cleaners are expensive (~4K) and often rely on clean air or oxygen to generate the required plasma. Our goal was to develop a cheaper working plasma generation system for less than $100, using parts easily obtainable from an average hardware store, and with no specialized tools. A low cost room-air microwave plasma generator was fabricated using a 400W, 2450MHz Sharp brand "Household Microwave Oven" for the purpose of exploring the possibility of irreversible sealing onto glass with minimal budgetary commitment. Irreversible sealing was observed via peel testing. Goniometry testing to verify hydrophilicity was performed on PDMS substrates treated with the room-air microwave plasma system developed here, and was compared with the commercially available Harrick Plasma (PDC-32G) plasma cleaner.

387776 One-Step Cell Lysis and DNA Concentration in a Multisection Insulator-Based Dielectrophoretic Device

Abstract:

The present work describes an application of a BioMEMS device for lysis of *E. coli* and the concentration of plasmidic DNA. The design of the microchannel fabricated in PDMS includes two sections of insulators, and DC electric fields are generated along its length to continuously produce irreversible electroporation on the cells and concentrate plasmidic DNA by negative dielectrophoresis. It also allows the redirection of trapped and concentrated particles to specific outlets to aid in the recovery of the targeted analytes. This method can be used as an alternative to traditional chemical or physical procedures for DNA, ARN, or protein recovery from cells, in the aim of the miniaturization of laboratory processes.

DEP Isolation and Detection of Cancer Related DNA Biomarkers – A Comparison of PCR and DNA Sequencing Results for Blood and Plasma

Michael J. Heller
University of California San Diego, Department of Bioengineering, Department of Nanoengineering, La Jolla, CA, USA 92093-0448

The analysis of cell free circulating (cfc) DNA continues to become more widely used for cancer diagnostics and management. We have now demonstrated a dielectrophoretic (DEP) based approach that allows cfc-DNA and RNA biomarkers from chronic lymphocytic leukemia (CLL) patients to be isolated in about 10 minutes directly from a small volume (25ul) of patient whole blood or plasma. Cfc-DNA isolated from fifteen CLL patient blood samples and eleven CLL patient plasma samples was PCR amplified to identify the VHL genotype and then sequenced. The DNA sequencing results for ccf-DNA isolated by DEP were compared to two gold standard methods for CLL analysis. New DEP microarray devices specifically designed for the isolation of cfc-DNA from high conductance buffer and blood samples were obtained from Biological Dynamics (La Jolla, CA). DEP experiments were carried out by adding about 25 ul of blood sample to the DEP microarray device. The DEP field was applied at 10 kHz and 20 V_{p-p} for 3 minutes. The microarray was then washed with 0.5x PBS with the DEP field on and examined by epifluorescent microscopy. Cfc-DNA was then removed from device and PCR amplified using VHL patient specific primers. The figure below shows the fluorescent image results for ccf-DNA isolated by DEP from one normal blood sample (Normal-1) and three representative CLL blood samples (CLL-6, CLL-9, CLL-10). On the right are 3D fluorescence intensity images created by MATLAB, which provide better visualization of the relative amounts of ccf-DNA that were isolated. Overall, the fluorescent ccf-DNA levels were higher in most of the CLL patient samples when compared to the fluorescent DNA levels obtained from the normal blood samples. Elution of the ccf-DNA from the DEP device for quantification and PCR analysis required less than ten minutes. The PCR and DNA sequencing results for ccf-DNA isolated from 15 CLL patient blood samples by the DEP process were exactly comparable to results obtained using conventional methods, which take several hours to complete, are more laborious (many steps) and require much larger plasma/blood samples. The results for eight of eleven CLL plasma samples compared well with the gold standards. Generally, cfc-DNA for "liquid biopsies" is isolated form plasma using relatively long and involved processes which are not suitable for point of care (POC) diagnostics. The new DEP devices are now being used to collect cfc-DNA/RNA from other cancer (solid tumor) and TBI patient blood and plasma samples, setting the stage for "seamless sample to answer" diagnostics.

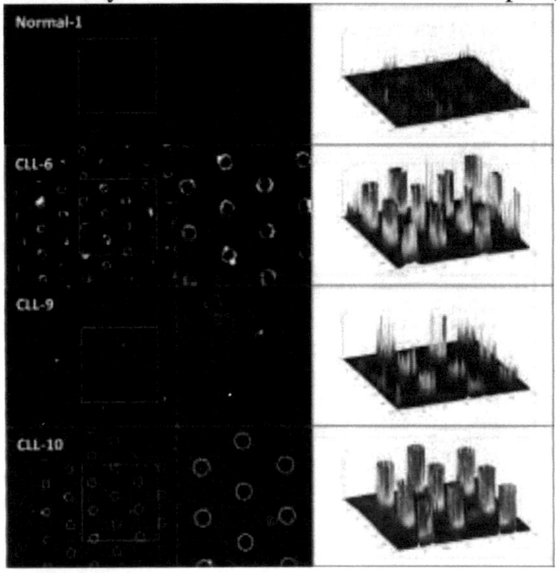

(1) Sonnenberg A, Marciniak JY, Rassenti L, Ghia EM, Skowronski EA, Manouchehri S, McCanna JP, Widhopf II GF, Kipps TJ and **Heller MJ**, "Dielectrophoretic Isolation and Detection of Cancer Related Circulating Cell Free DNA Biomarkers from Blood and Plasma", Electrophoresis, accepted for publication 2014; (2) Sonnenberg A, Marciniak JY, Rassenti L, Ghia EM, Skowronski EA, Manouchehri S, McCanna JP, Widhopf II GF, Kipps TJ and **Heller MJ**, "Rapid Electrokinetic Isolation of Cancer-Related Circulating Cell Free DNA Directly from Blood", Clinical Chemistry, 60:3, pp. 500-509, 2014.

376704 Exploiting Absolute Negative Mobility with Dielectrophoresis for Mitochondrial Sample Preparation

Abstract:

The study of mitochondria is important due to the organelle's contribution to many cellular functions, including metabolism, cellular signaling and apoptosis. The atypically sized giant mitochondria[1] were found in the models of aging and some diseases. Thus it would be advantageous for related studies if mitochondrial subpopulations could be prepared in accordance with their particular sizes. Conventional methods such as centrifugation and electrophoresis can separate mitochondria from cell cytoplasm and further purify them from other organelles. However, the separation of mitochondrial subpopulations has rarely been studied. Here, we present the numerical simulation with the design of a microfluidic device to elucidate absolute negative mobility (ANM) utilizing insulator-based dielectrophoresis (iDEP) for the separation of mitochondrial subpopulations.

ANM has recently been proposed as a separation mechanism for micron-sized particles[2]. It is the motion of a particle into the direction opposite to a net applied force[3], which can be realized when Brownian particles migrate in a non-linear structure under non-equilibrium conditions. For instance, when particles migrate with flow (y direction in Figure 1a) in an ANM structure, they could be trapped by adjacent posts which form a gap smaller than their diameter. On the other hand, probability exists that they can also avoid the trap due to their diffusion in the direction perpendicular to the flow. This probability for a Brownian particle - being trapped or avoiding a trap - relates to the diffusion property of the particle, i.e. its size. Therefore, under non-equilibrium conditions particles within a certain size range may move towards the direction of a net force while other particles may move in the opposite direction and consequently, separation based on ANM may be realized. The design of the ANM structure has been studied previously both in theory with the variation of geometric post shapes and in experiment with micron-sized colloidal particles in a microfluidic device[4]. However, this approach is not transferrable to evoke ANM for the separation of mitochondrial subpopulations with typical sub-micron sizes (0.1 - 1 μm)[1]. Thus, we additionally exploit the dielectrophoretic properties of this important organelle in tailored microdevices in order to generate ANM. DEP is a migration phenomenon based on the polarization property of a particle in an inhomogeneous electric field. This property has been used to separate, concentrate or distinguish many biological species, including cells, tissue, proteins, bacteria and DNA[5]. Our previous studies[6] revealed that isolated Fischer 344 rat semimembranosus muscle mitochondria and C57BL/6 mouse hepatic mitochondria exhibited negative DEP (nDEP) under both DC (0 - 3000 V/cm) and AC (0 - 2000 V/cm, 0 - 50 kHz) conditions in low conductivity buffer in physiological pH. In addition, the potential thresholds for trapping the mitochondrial samples were weakly depended on the applied frequencies. Based on that, a structure modified to suit the small size exploiting the nDEP property of mitochondria was designed (Figure 1b).

First, the migration properties of particles exhibiting nDEP were studied in numerical simulation using *COMSOL Multiphysics 4.4*. Properties of micron and sub-micron polystyrene beads were used to mimic the mitochondrial subpopulations. Non-equilibrium conditions were realized by an alternating electric field, which was tunable by combining the static study result of the *Electric Currents* module with a rectangle and an analytic function. This was further coupled with the time-dependent study using the *Creeping Flow* module, resulting in an alternating flow profile, which was further integrated with DEP and Brownian forces in the time-dependent study of the *Particle Tracing for Fluid Flow* module. The resultant particle trajectory allowed for the investigation of the conditions such as the applied potential and the dimensions of the posts and the gaps needed for realizing ANM for the separation of micron- and sub-micron sized particles and thus accessing a regime suitable for mitochondrial subpopulations. We found that with certain conditions, ANM could be observed with particles of 5-μm diameter (Figure 1d) while 1-μm particles do not show ANM. This novel approach to induce differential ANM migration behavior for particles of varying sizes allows retaining simple photolithographic device fabrication strategies for experimental ANM applications with organelles, and similarly preserves the size selectivity for sub-micron bio-species.

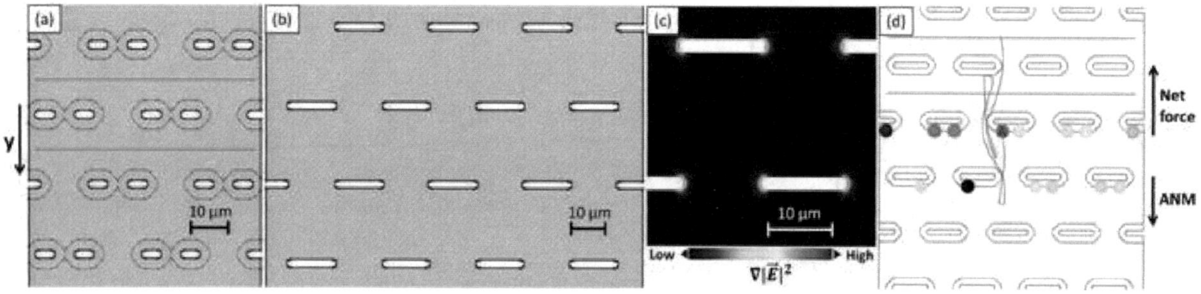

Figure 1: Schematics of the ANM devices. Electric potential was applied along the y-axis. The posts and the channel walls are outlined in blue, while the black lines in (a) indicate the virtual walls which are used to exclude particles from the post regions with a distance equal to or smaller than their radius. (a) The two black horizontal lines indicate the positions where particles are released in the simulation. Because the small gap is 5 μm wide, this exemplary structure is suitable for the ANM study of particles with diameter no smaller than 5 μm. (b) Alternative layout to couple a DEP component. This structure is suitable for the ANM study of mitochondria of which the typical size range is from 0.1 μm to up to 1 μm[1]. (c) Zoom-in of (b) showing the electric field gradient distribution around the posts. If there is flow in the y direction, a particle exhibiting nDEP will be pushed towards and trapped in the center of a post similarly to a geometrical trap as shown in (a). (d) ANM behavior simulated with particles of 5 μm diameter in two consecutive time periods of the applied frequency. Particles started at the two horizontal lines and migrated with flow in the y direction. The trajectory of particle A was labeled in the figure demonstrating ANM.

References

1. M. Navratil, A. Terman, and E. A. Arriaga, Exp. Cell Res. **314,** 164 (2008).

2. A. Ros, R. Eichhorn, J. Regtmeier, T. Duong, P. Reimann, and D. Anselmetti, Nature **436,** 928 (2005).

3. R. Eichhorn and P. Reimann, Acta Physica Polonica B **37,** 1491 (2006).

4. J. Regtmeier, S. Grauwin, R. Eichhorn, P. Reimann, D. Anselmetti, and A. Ros, Journal of Separation Science **30,** 1461 (2007).

5. T. Z. Jubery, S. K. Srivastava, and P. Dutta, Electrophoresis **35,** 691 (2014).

6. J. Luo, B. G. Abdallah, G. G. Wolken, E. A. Arriaga, and A. Ros, Biomicrofluidics **8,** 021801 (2014).

Carbon-electrode Dielectrophoresis for Sample Preparation

R. Martinez-Duarte

Department of Mechanical Engineering, Clemson University,
Clemson, South Carolina 29634, USA

This work presents the latest applications of carbon-electrode Dielectrophoresis for sample preparation. The use of 3D carbon electrodes allows for higher throughput and efficiency when compared to more traditional DEP techniques, such as metal-electrode and most forms of insulator-based DEP.

The applications presented here all feature a protocol that allows for enrichment and purification of a targeted particle population from a sample; and effectively enables sample preparation for downstream analysis. Such protocol features three main stages: 1) extract targeted particles from the sample by trapping them on the electrode array using pDEP; 2) hold the particles on the electrode array by pDEP forces while a washing solution is flowed through the channel; and 3) release the targeted particles, by de-polarizing the electrode array, for collection at the exit of the channel. In such way, one can enrich and purify a selected particle population from a given sample. Two applications are presented: a) elimination of natural contaminants to increase sensitivity of PCR-based protocols; and b) enrichment of bacterial persisters from an antibiotic-treated sample.

The final goal is the integration of independent carbon-electrode arrays, optimized in their geometry and polarizing signals, to achieve an automated sample preparation device comprising functions such as concentration, purification and lysis.

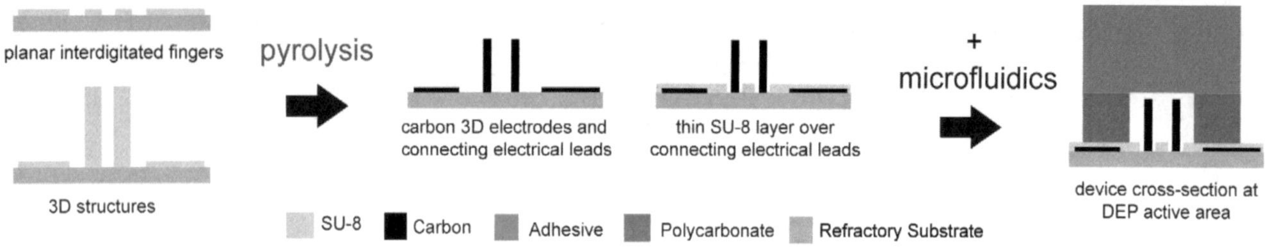

Figure 1. Overview of the fabrication of carbon electrodes

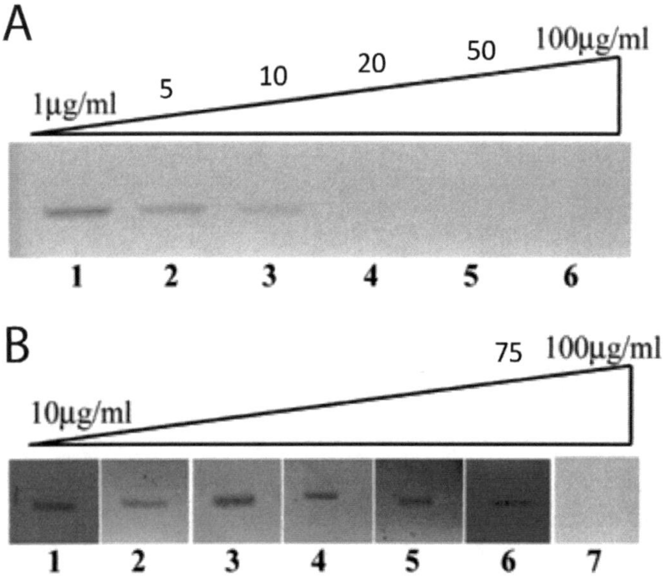

Figure 2. PCR results demonstrating A) the inability of only PCR analysis to clearly detect yeast cells when the concentration of humic acids in the sample is above 10 ug/ml and B) significant improvement obtained when using a carbonDEP-based sample preparation module before PCR analysis: humic acids concentrations up to 75 ug/ml are effectively removed from the sample to enable yeast detection.

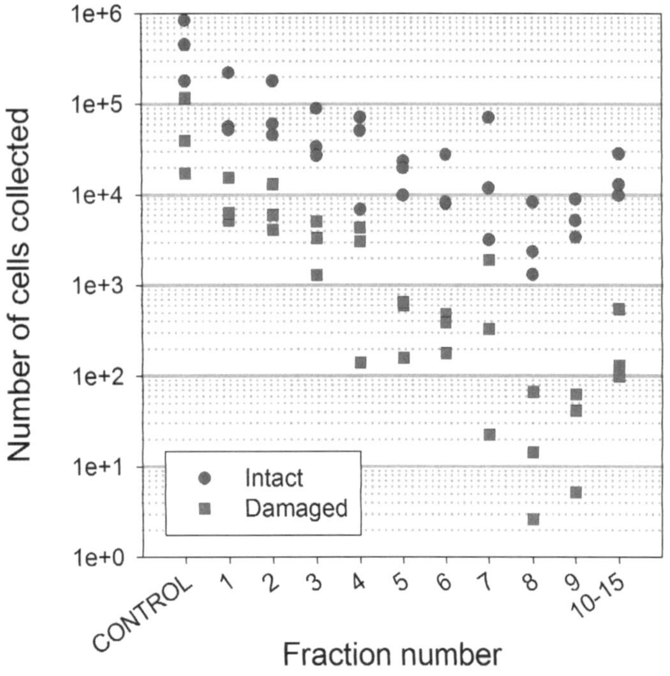

Figure 3. Intact and damaged cell counts (y-axis) of the sequential fractions recovered during the carbonDEP assay. The control fractions refers to the original sample, flowed through the chip without polarizing the electrode array. Green triangles represent viable cells while inverted red triangles represent non-viable cells. Symbols correspond to three representative experiments.

Rapid, Specific, and Efficient Affinity Purification of Target Molecules by Combining Isotachophoresis and Affinity Chromatography

Viktor Shkolnikov and Juan G. Santiago
Stanford University
Nov. 18, 2014

Background

Affinity chromatography

- Commonly used to purify proteins, nucleic acids
- Advantages:
  - highly specific
  - binary (eluted/not eluted)
  - direct stringency control via wash
- Disadvantages:
  - often slow kinetics (10s of hours)
  - samples with large matrix component cause fouling

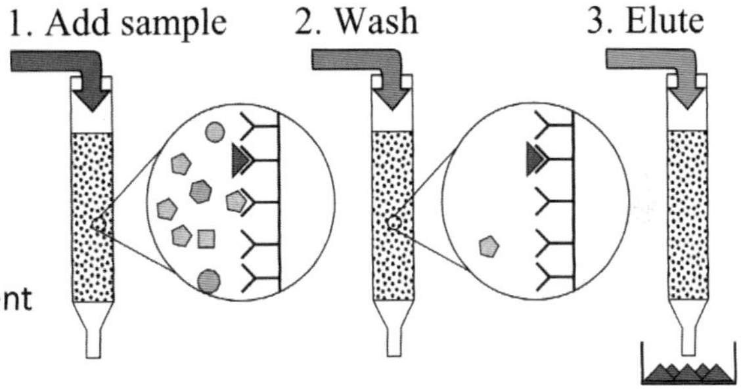

Isotachophoresis

- Electrokinetic technique used to preconcentrate and separate analytes
- Offers tight control over analyte concentration and position

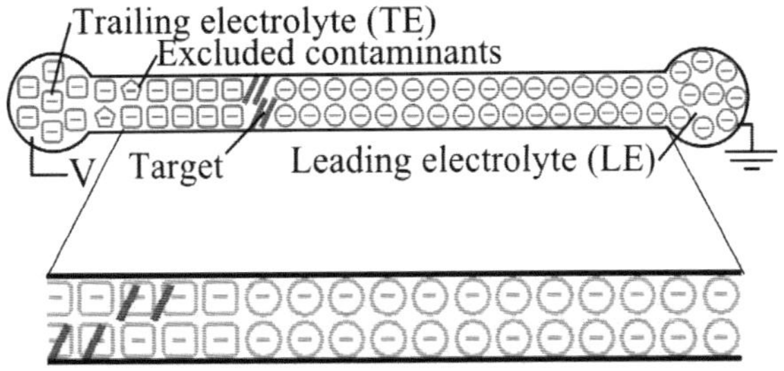

Coupling isotachophoresis and affinity chromatography

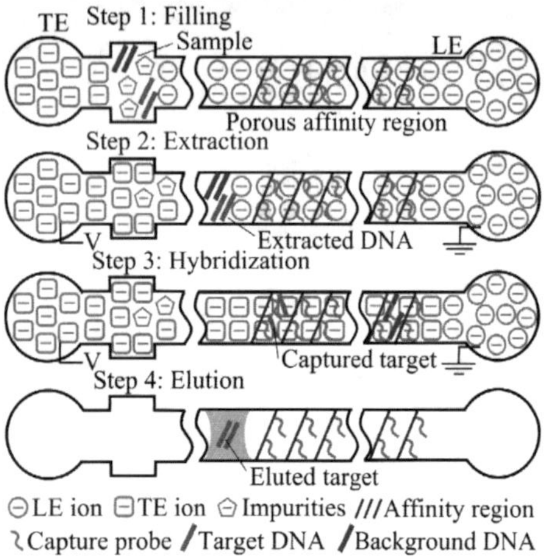

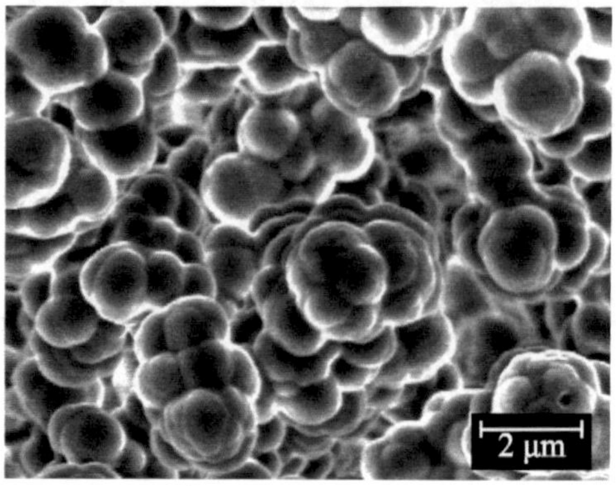

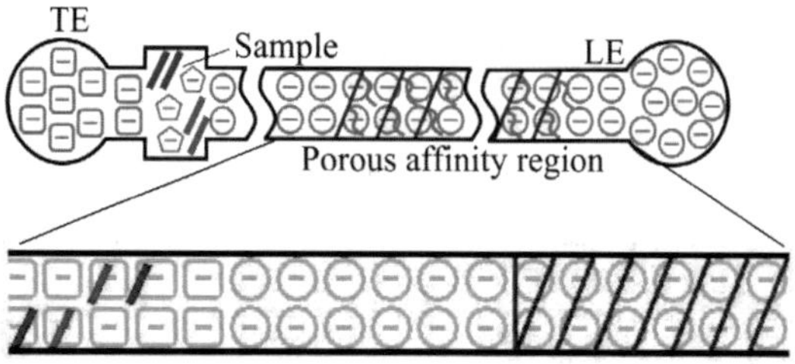

Model for coupling isotachophoresis and affinity chromatography

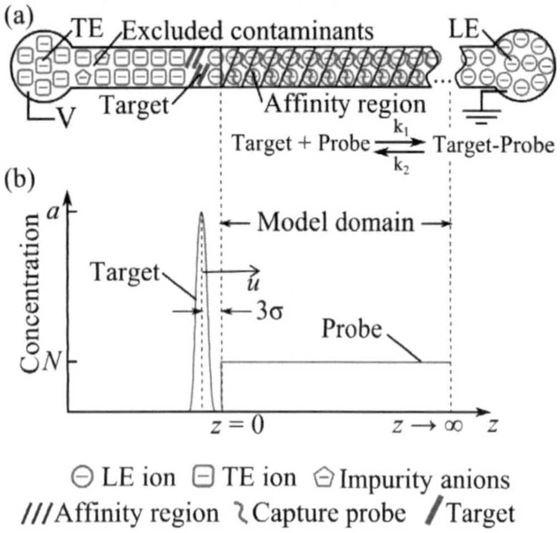

Shkolnikov & Santiago, Part I: Theory, 83, 13, *Anal Chem.*, 2014
Shkolnikov & Santiago, Part II: Experimental Study, 83, 13, *Anal Chem.*, 2014

Model for coupling isotachophoresis and affinity chromatography

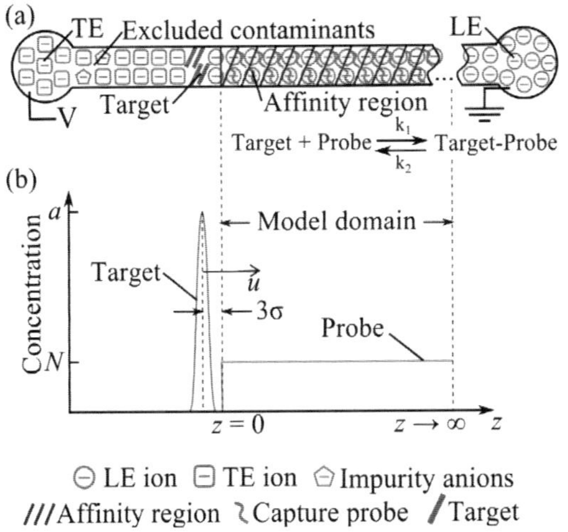

Nondimensionalization of variables:

$c^* = \dfrac{c}{N}$ dimensionless target concentration

$n^* = \dfrac{n}{N}$ dimensionless target-probe complex conc.

$t^* = tk_1N$ dimensionless time

$z^* = \dfrac{z}{u/k_1N}$ dimensionless axial distance from leading edge

---

$\beta = \dfrac{k_2}{k_1N}$ dimensionless dissociation constant; reversibility of reaction

$\alpha = \dfrac{a}{N}\sqrt{2\pi}$ dimensionless peak target concentration ;

$Da = \dfrac{\sigma k_1 N}{u}$ dimensionless target distribution width; advection / reaction time scales

αDa dimensionless total target amount; local column saturation

Shkolnikov & Santiago, Part I: Theory, 83, 13, *Anal Chem.*, 2014
Shkolnikov & Santiago, Part II: Experimental Study, 83, 13, *Anal Chem.*, 2014

Model for coupling isotachophoresis and affinity chromatography

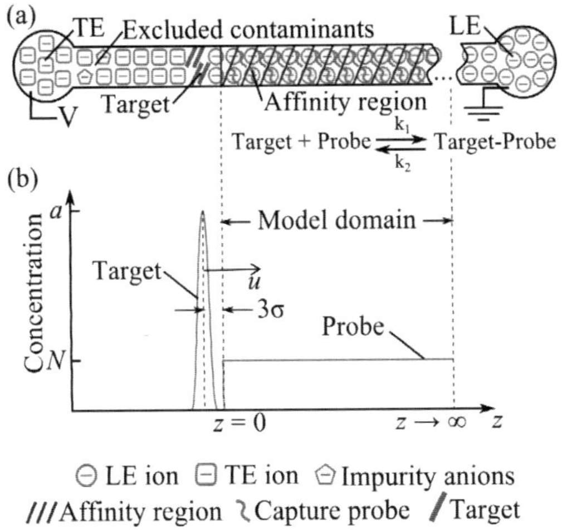

Nondimensionalization of variables:

$c^* = \dfrac{c}{N}$ dimensionless target concentration

$n^* = \dfrac{n}{N}$ dimensionless target-probe complex conc.

$t^* = tk_1N$ dimensionless time

$z^* = \dfrac{z}{u/k_1N}$ dimensionless axial distance from leading edge

---

$\beta = \dfrac{k_2}{k_1N}$ dimensionless dissociation constant; reversibility of reaction

$\alpha = \dfrac{a}{N}\sqrt{2\pi}$ dimensionless peak target concentration ;

$Da = \dfrac{\sigma k_1 N}{u}$ dimensionless target distribution width; advection / reaction time scales

αDa dimensionless total target amount; local column saturation

Transport and reaction of target:

$$\frac{\partial c^*}{\partial t^*} + \frac{\partial c^*}{\partial z^*} + \varepsilon \frac{\partial(\tilde{u}^* c^*)}{\partial z^*} = \frac{Dk_1N}{\tau^2 u^2}\frac{\partial^2 c^*}{\partial z^{*2}} - c^*(1-n^*) + \beta n^*$$

$$\varepsilon = (\mu_{a,TE} - \mu_{TEion,TE})/\mu_{TEion,TE}$$

Reaction of immobilized probe:

$$\frac{\partial n^*}{\partial t^*} - c^*(1-n^*) + \beta n^* = 0$$

Boundary condition for target:

$$c^*(0,t^*) = \frac{a}{N}\exp\left[-\left(\frac{ut^*}{k_1N\sqrt{2}\sigma} - \frac{3}{\sqrt{2}}\right)^2\right]$$

Initial condition for target:

$$c^*(z^*,0) = 0$$

Initial condition for probe:

$$n^*(z^*,0) = 0$$

Shkolnikov & Santiago, Part I: Theory, 83, 13, *Anal Chem.*, 2014
Shkolnikov & Santiago, Part II: Experimental Study, 83, 13, *Anal Chem.*, 2014

Model predictions: capture length and capture time

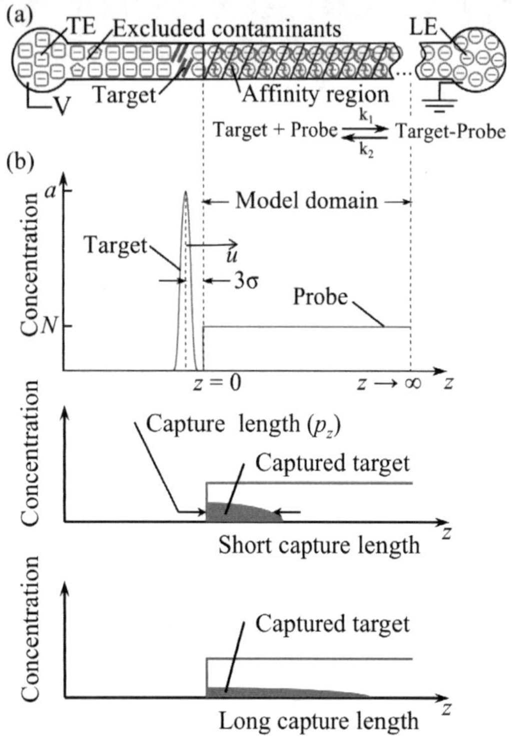

Shkolnikov & Santiago, Part I: Theory, 83, 13, *Anal Chem.*, 2014
Shkolnikov & Santiago, Part II: Experimental Study, 83, 13, *Anal Chem.*, 2014

Model predictions: capture length and capture time

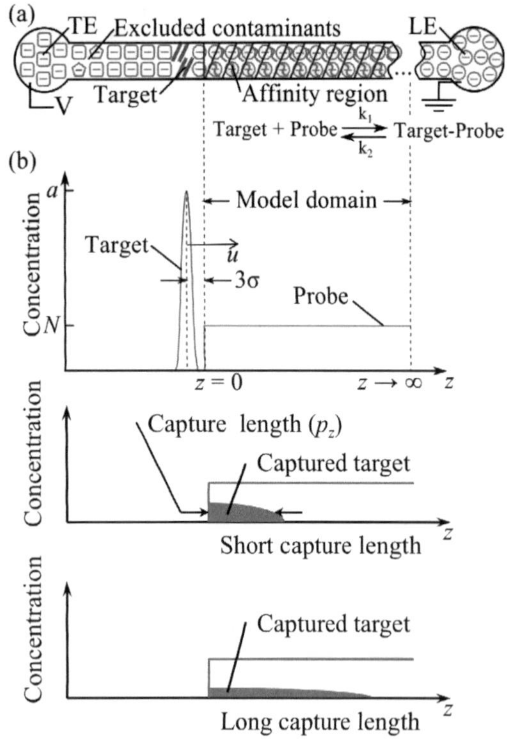

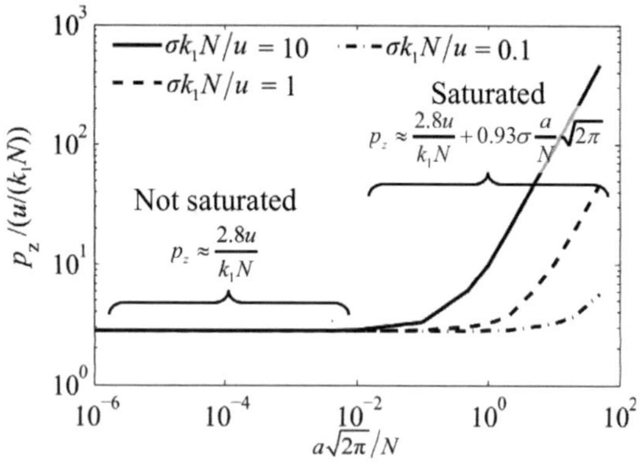

Shkolnikov & Santiago, Part I: Theory, 83, 13, *Anal Chem.*, 2014
Shkolnikov & Santiago, Part II: Experimental Study, 83, 13, *Anal Chem.*, 2014

Model predictions: capture length and capture time

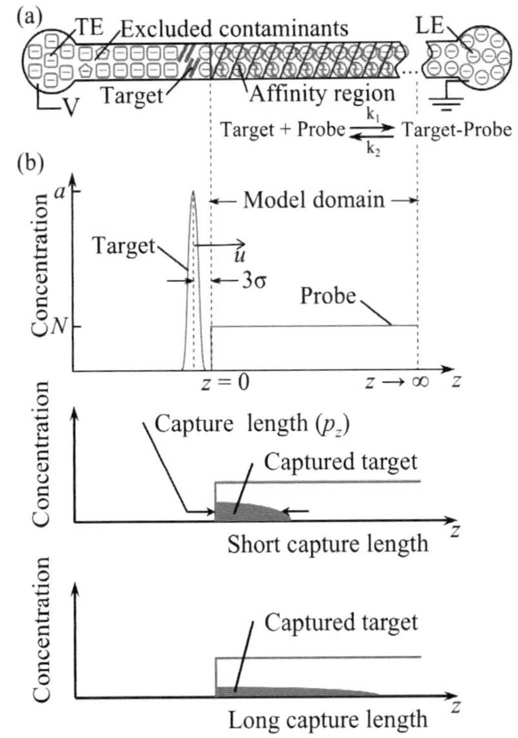

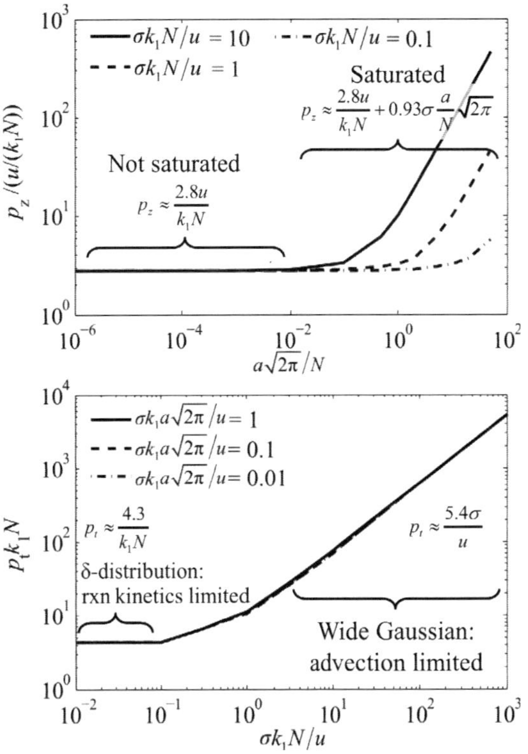

Shkolnikov & Santiago, Part I: Theory, 83, 13, *Anal Chem.*, 2014
Shkolnikov & Santiago, Part II: Experimental Study, 83, 13, *Anal Chem.*, 2014

Experimental validation of the model

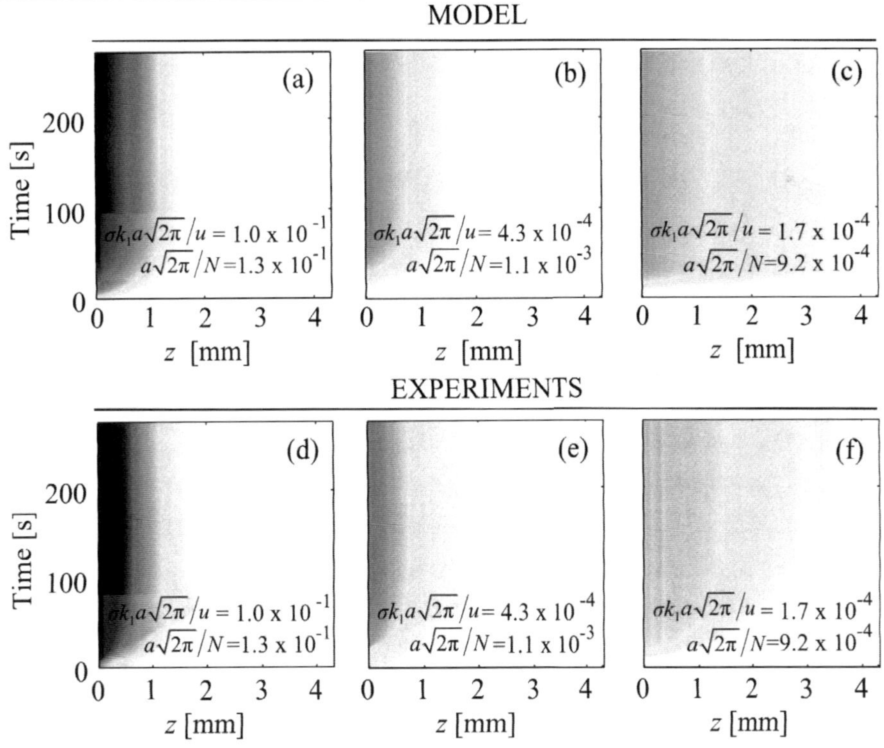

Shkolnikov & Santiago, Part I: Theory, 83, 13, *Anal Chem.*, 2014
Shkolnikov & Santiago, Part II: Experimental Study, 83, 13, *Anal Chem.*, 2014

Experimental validation of the model

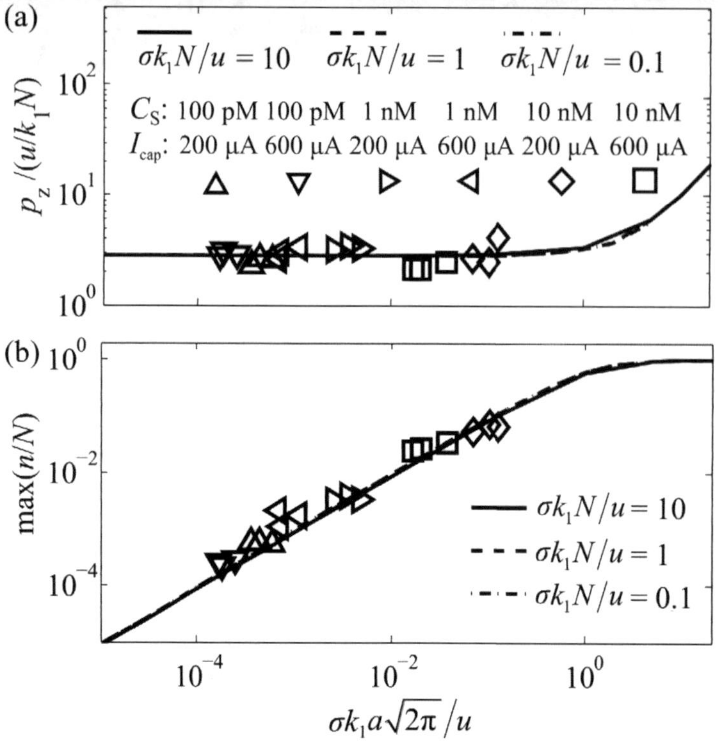

Shkolnikov & Santiago Part I: Theory, 83, 13, *Anal Chem.*, 2014
Shkolnikov & Santiago, Part II: Experimental Study, 83, 13, *Anal Chem.*, 2014

Sequence specific purification of rare target DNA from 10,000x background DNA

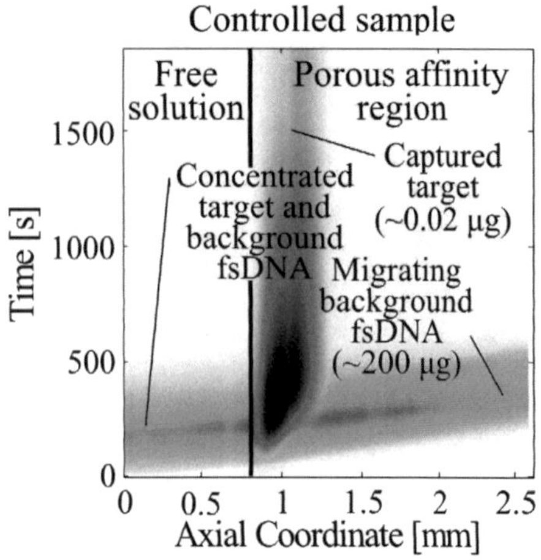

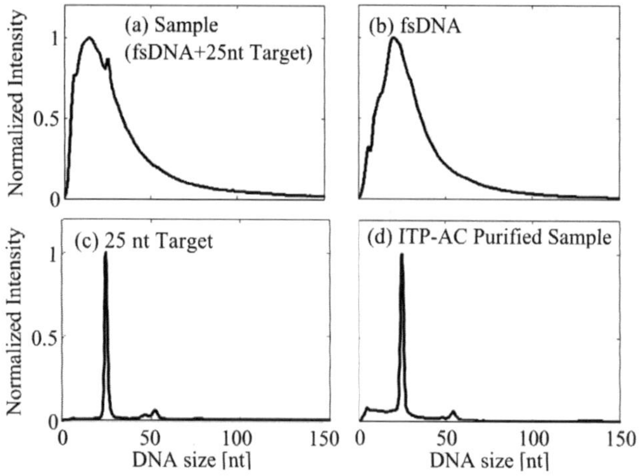

- Sample: 25 mM HEPES, 50 mM Tris, fish sperm DNA, target DNA.
- Eluted with 50 mM NaOH (quenched to pH 7 with 200 mM HEPES)
- Analyzed eluted sample with Agilent BioAnalyzer.
- Purification in under 1 mm column length and in less than 500 s.
- Separation resolution increases linearly with time.

Shkolnikov & Santiago Part I: Theory, 83, 13, *Anal Chem.*, 2014
Shkolnikov & Santiago, Part II: Experimental Study, 83, 13, *Anal Chem.*, 2014

Purification of rare (13,000x) DNA target from whole human blood lysate

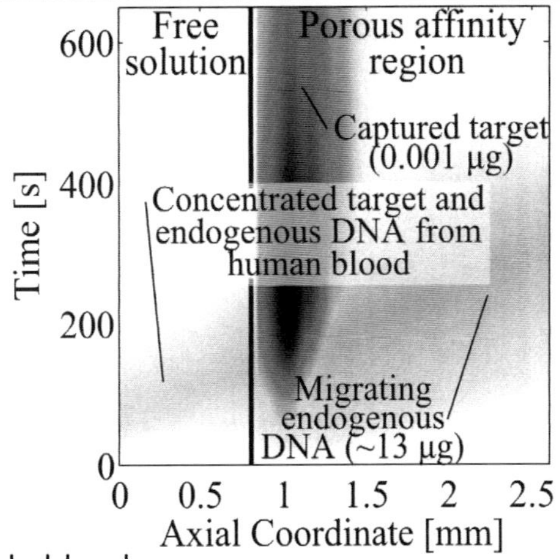

- Sample: 12.5 μl of whole blood
- Prepared sample (25 μl):
  - 12.5 μl of whole blood
  - 1% Triton-X100 } For leukocyte lysing
  - 0.2 mg/μl of Proteinase K }
  - 1 ng of 25 nt Cy5 target DNA } For DNA visualization
  - 0.1x SYBR Green I }
- Separation in under 500 s and in under 1 mm column length

Shkolnikov and Santiago, unpublished result, 2014

Summary

- Coupling isotachophoresis and affinity chromatography:
  - improves column utilization
  - decreases purification time
  - prevents column fouling
  - obviates the need for specialized high pressure pumps
- We developed an analytical model for ITP-AC
  - collapses 6 key process parameters to 3 non-dimensional parameters
  - yields simple analytical relations for capture length and capture time
  - experimentally validated over a wide range of parameters
- We demonstrated ITP-AC for
  - sequence specific purification of rare target from > 10,000x DNA background
  - from a complex, clinically relevant sample (whole human blood)
  - with few manuals steps

NSF BIO-X NATIONAL INSTITUTES OF HEALTH

385839 Designing an Integrated Biosensing Platform for Sample-to-Answer Solution

Abstract:

Real time PCR (rt-PCR), the current gold standard for pathogen detection, is expensive and requires trained personnel/bulky instrumentation making it impractical for field applications. We envision to develop a push-button sample-to-answer low-cost, self-powered and portable instrument with a slot where an integrated microfluidic chip is inserted for detection of any target pathogen of interest. The objective is to use our patented nanomembrane-based biosensing technology to develop a low-cost, rapid (45 min), specific (single base mismatch) and sensitive (1 fM) pathogen detection platform without any PCR amplification. The integrated microfluidic biochip consists of a sample pretreatment unit to isolate nucleic acid from the target sample and nanomembrane-based molecular sensing/pre-concentration unit.

The sample pretreatment unit is fabricated based on gel electrophoresis where an agar gel is used to separate sample reservoir from microfluidic channel. The DC field is applied to draw the negatively charged nucleic acid molecules to a microfluidic channel from chemically lysed target sample. This allows to filter out the negatively charged biomolecules from the cell debris. This sample preparation is then integrated with an ion-selective nanomembrane based preconcentration unit to concentrate the isolated nucleic acid at a specific position in a microfluidic channel and biosensor is placed in that location for detection of target pathogen of interest. Several interesting phenomena are known to take place at the interface of an ion selective environment (nanochannel or ion exchange membrane or particle) and the surrounding electrolyte in DC or AC electric fields. These phenomena include formation of depletion and concentration regions around the ion selective nanoporous materials, development of a space charge region, development of electrically driven vortices, water splitting reaction etc. The concentration of biomolecules in our integrated platform is due to the depletion phenomenon developed at the cation-exchanged membrane interface and application of pressure driven flow in the opposite direction of depletion front. The preconcentration makes the biosensing rapid and sensitive by reducing the diffusion time and also concentrating the target molecules by 200x. The anion-exchanged nanomembrane based biosensing is based on a charge inversion phenomenon that occurs when negatively charged DNA/RNA molecules hybridize with specific probes attached to nanoporous materials bearing a positive fixed charge. This is detected by measuring a change in current-voltage characteristics of the system before and after hybridization. Instead of the classical electrochemical sensing technique, we utilize a much more sensitive conductance signature based on sensitivity of ion current across the nanoporous membrane to surface charge density change effected by hybridized molecules. The sensitivity involves a nonlinear correction to the conductance and is hence much more sensitive than EIS or other linear electrochemical sensing techniques.

391732 An Entropic Force Microscope Enables Nano-Scale Conformational Probing of Biomolecules

Abstract:

Entropy is fundamentally important in nearly all physical phenomena, but nature's exquisite control over the underlying randomness masks a complexity that has proven challenging to synthetically replicate. Here we show how the inherently disordered dynamics governing macromolecular transport in confined surroundings, conventionally suppressed as an undesirable noisy background, can paradoxically be precisely controlled. This ability lays a foundation for an entropic force microscope, a sensitive probe of nanoscale biomolecular conformation capable of resolving previously unseen details about DNA-protein binding interactions at size scales below the limits of conventional techniques.

Microfluidic Quantification of Transcriptome and Functional Proteins from the Same Single Cells

Jun Wang*, James Heath**
*University at Albany, The State University of New York, USA
**California Institute of Technology, USA

Transcriptomic analysis can powerfully reveal the regulatory mechanism of cells under various conditions. With the dramatic advancement of next-generation sequencing technologies and microfluidics, the analysis of single-cell transcriptome becomes possible for biomedical researchers. However, the interpretation of gene expression information is tricky, since cell functions, such as differentiation, migration, and communication, may not be corresponding to the intracellular mRNA levels. That is because mRNAs are usually intermediate products for translation, and proteins take most roles for cell functions. Although protein networks and transcript networks are relatively far away in cell regulatory machinery, we are interested in how those networks interact with each other, particularly how transcriptional networks control cell functions represented by protein secretion.

We developed a proof-of-concept approach to measure protein secretion by a microchip and gene expression by qPCR or RNA sequencing (Figure 1). The single-cell barcode chip (SCBC) was used to analyze secretion profile of CD8+ T cells. We incubated cells in sealed microchambers first. In order to keep cells in microchambers after on-chip incubation, each chamber was designed to have a dent at the center, large enough to allow a cell to reside in. A barcode slide was mated with the PDMS replica, and the chip was mounted on a clamp. After incubation, the PDMS part was carefully separated from the barcode slide, in order to retain cells in microchambers. The PDMS part was kept at low temperature until the completion of sandwich ELISA on the barcode slide. The whole barcode slide was scanned and analyzed by customer programs. The address of each cell microchamber with one cell residing there was recorded as an index for further pickup. A home-made stage was built to stabilize a syringe connected with a needle (Hamilton; 32 gauge), which was used to draw single cells and transfer them to a 96 well plate. The single cells were lysed before analysis by qPCR using Fluidigm BioMark. Alternatively we also sequenced mRNAs by Illumina 2500 using a protocol developed by our group.

We first demonstrated the ability of quantitating both proteins and mRNAs simultaneously on CD8+ T cells from a melanoma patient (Figure 2). Cytokine secretion is the major function of T cells. In clinical adoptive immunotherapy, engineered T cells are supposed to attack cancer cells in patients' body. However, recently we found there are always only a small portion of T cells functioning properly through cytokine secretion, and no biomarkers were discovered yet to distinguish functional T cells from the rest. Through analysis of transcriptome, we found the potential biomarkers at the transcriptional level and transcriptional networks that are correlated with the secretion of cytokines. Based on weighted gene co-expression network (WGCNA) analysis, we searched the genome-level transcripts of simulated T cells with various durations, and discovered the cytokine-secretion related sub-networks that can be targeted to improve T cell engineering in the future (Figure 3).

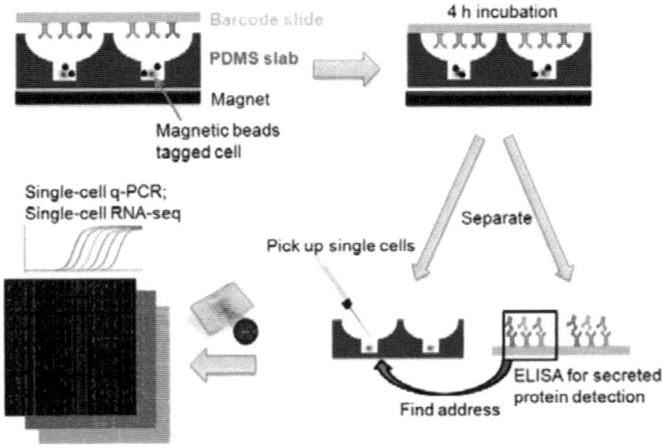

Figure 1: Procedure of quantitating both secreted proteins and mRNAs from the same single cells. Single cells tagged with magnetic beads are incubated in the microchambers of a single-cell barcode chip for hours. Proteins of interest that are secreted will be captured by the antibody array in each chamber. After separation of the PDMS slab and the barcode slide, sandwich ELISA is performed on the barcode slide. The corresponding microchamber address of secreting cells can be found on the PDMS slab, and the single cell is picked up for further analysis by either single-cell qPCR or RNA sequencing.

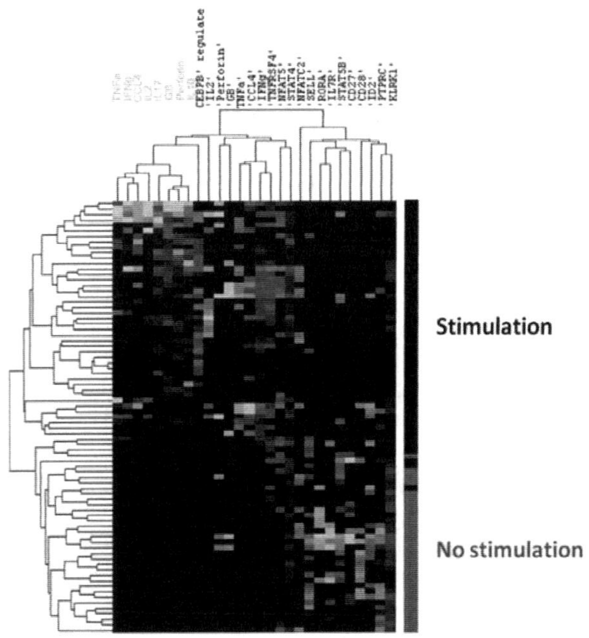

Figure 2: Separation of T cell subpopulations and their gene expression patterns according to cell functions. Secretion profiles (in turquoise color) of single cells are clustered and showed by the heatmap, where yellow color represents high expression level, and dark color represents low expression level. Expression of genes (right; black font) is also presented in the same graph. Data of stimulated T cells and those without stimulation are marked in the right.

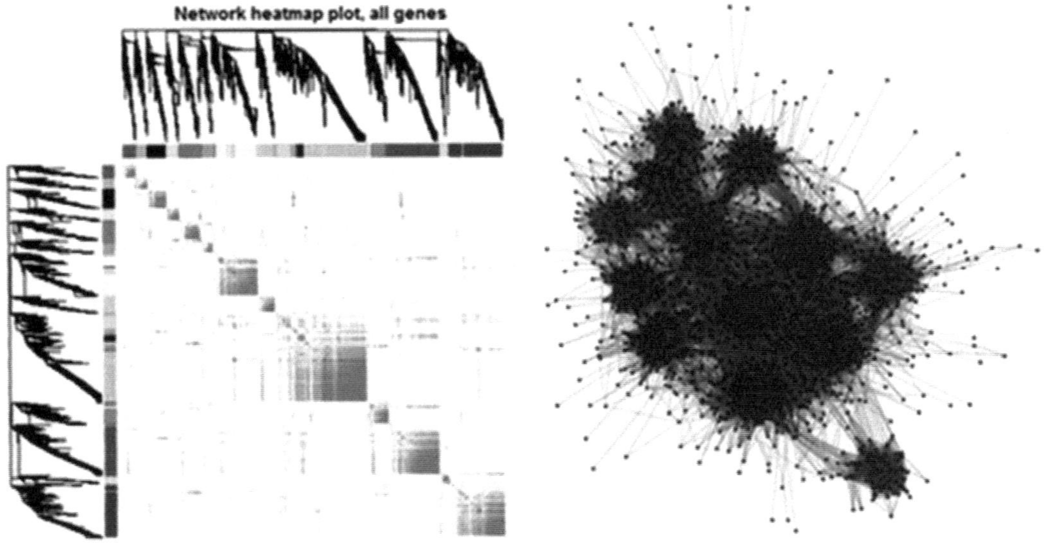

Figure 3: Construction of single-cell transcriptional networks using weighted gene co-expression network (WGCNA) analysis.

385691 Probing Space Charge and Resolving Overlimiting Current Mechanisms at the Micro-Nanochannel Interface Using Electrochemical Impedance Spectroscopy

Abstract:

We present new experimental and theoretical results concerning the nature of the overlimiting current in micro-nanochannel devices. Previous work has focused on the respective roles of a variety including electro-convective instability, shock-propagation, and electro-osmosis. Here we demonstrate a primary mechanism for the overlimiting conductance arises from the structure of the extended space charge developing at the depleted interface during high-current concentration polarization. This results in a maximum of the resistance which is captured well by asymptotic analysis of the extended space charge. Further, we show that this effect is strongly modulated by voltage-dependent changes in selectivity and the field-focusing effect. Lastly, we present experimental results demonstrating an anomalous resistance minimum occurring in devices with shorter nanochannels. Based on results of numerical simulation, this is hypothesized to result from the interaction between surface conduction mechanisms and the space charge structure at the nanochannel entrances.

386800 An Orbital Shear Platform for in-Vitro Real-Time Endothelium Characterization

Abstract:

Introduction

Atherosclerosis is the buildup of plaque in lesions in the inner most lining of arteries, known as the endothelium. As plaque hardens, the arteries become narrow and limit blood flow leading to the onset of cardio disease, such as congestive heart failure, stroke and hypertension. Studies have shown that the development of atherosclerosis is linked to changes in the endothelium function [1-2]. Shear stress can control endothelium and in some cases leading to increased endothelium permeability which in turn can lead to lesions where plaques forms [3-4].

Electrical impedance measurements can be used to monitor the permeability and cell movement of adherent cells, such as endothelial cells [5]. In this work, we present an impedance based *in-vitro* platform used to monitor the behavior of cultured Human Umbilical Vein Endothelial Cells (HUVECs) under fluid-dynamic oscillatory shear.

Material and methods

The platform (Figure 1) consists of a microfabricated array of 200 μm (in diameter) gold sensing electrodes and complimentary counter electrodes on a borosilicate substrate. The chip contains 3 rows of electrodes which are spaced 120 degrees from each other. Electrodes are positioned at defined radial locations with respect to the center of the borosilicate chip. A PDMS mold forms a 35 mm diameter well. Shear stress is induced by placing the platform on an orbital shaker (Figure 2). Within the rotating orbital well, cells experience a range of shear stress values and in differing directions at different radial positions [6].

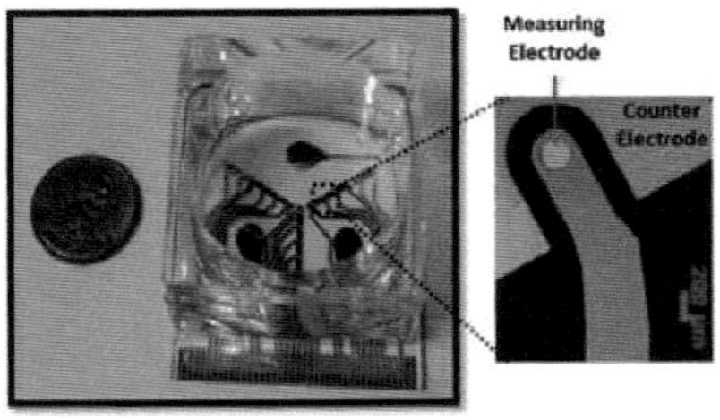

Figure 1: (Left) An image of the completed chip including a 35 mm PDMS well. (Right) A microscopic image of a sensing electrode (200 μm in diameter).

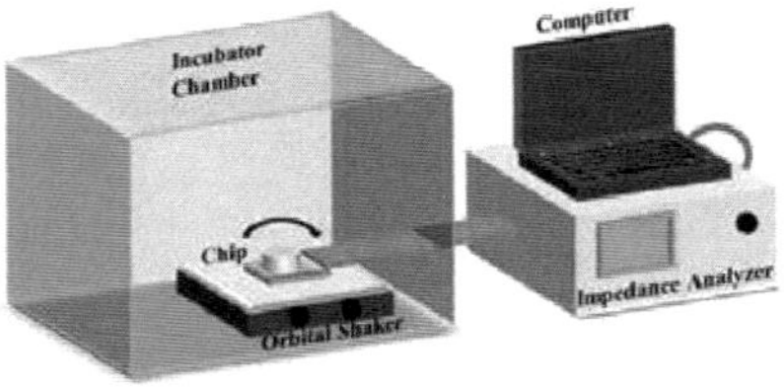

Figure 2: Illustration of the orbital shear *in-vitro* platform and its experimental setup.

For this work, HUVECs were seeded and allowed to adhere in the platform overnight in static condition. Impedance spectra (Agilent 4294A) were acquired one hour prior to the onset of shear and then 60 hours after initiating shear. Cells were monitored at radial positions 0, 3, 9, and 15 mm. The orbital shaker was set to 120 rpm and cells were exposed to ~0-10 dyne/cm^2 shear. Numerical simulations for this case are demonstrated in Figure 3.

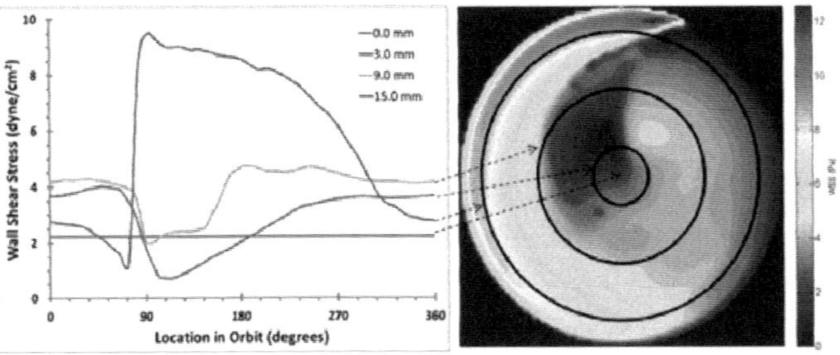

Figure 3: Numerical simulation of the shear magnitude at different radial position for a rotating cell culture well at 120 rpm, 0.95 cm orbital radius, and 2.0 mm fluid height.

Results and Conclusions

Acquired impedance measurements (Figure 4) indicate that at radial locations of 9 mm and 15 mm, impedance initially increased after initiating shear before reaching a plateau and subsequently decreasing. A previous publication suggests that the original increase is the result of cells flattening when fluid flow begins. While the subsequent decrease, demonstrates how cell junctions begin to weaken in order to align themselves in the direction of flow [7]. However, cells near the center at radial positions 0 mm and 3 mm were not uniformly oriented, similar to those observed in unsteady flow. Hence, an insignificant increase in impedance is observed which is indicative of a permeable endothelium.

The *in-vitro* platform presented within here is unlike other common *in-vitro* platforms based on microchannels because this device subjects cells to both unsteady and steady hydrodynamic shear simultaneously. Shear stress can be easily modulated by changing the rotational velocity and/or liquid volume, and the device requires very simple cell seeding and maintenance procedures. It also provides to acquire real time data, avoiding the need of lengthy fluorescent cell staining and microscope imaging.

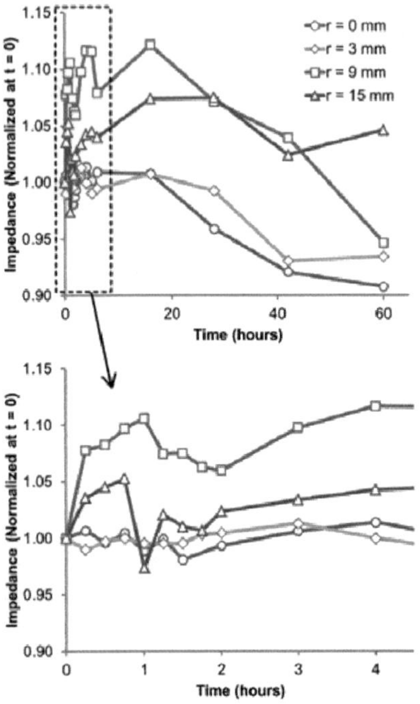

Figure 4: Normalized impedance measurements of HUVECs under orbital shear at various radial locations within the *in-vitro* platform.

REFERENCES

Davies, P.F. and S.C. Tripathi, *Mechanical-stress mechanisms and the cell – an endothelial paradigm.* Circulation Research, 1993. 72(2): p. 239-245.

Ross, R., *The pathogenesis of atherosclerosis - a perspective for the 1990s.* Nature, 1993. 362(6423): p. 801-809.

Keynton, R.S., et al., *The effect of graft caliber upon wall shear within in vivo distal vascular anastomoses.* J Biomech Engr., Feb. 1999, vol. 121, p. 79-87.

Davies, P.F., et al., *Turbulent fluid shear stress induces vascular endothelial cell turnover in vitro.* Proc. of the National Academy of Sciences, 1986. 83(7): p. 2114-2117.

Giaever, I. and C.R. Keese, *Micromotion of mammalian-cells measured electrically.* Proceedings of the National Academy of Sciences of the United States of America, 1991. 88(17): p. 7896-7900.

Chakraborty, A., et al., *Effects of biaxial oscillatory shear stress on endothelial cell proliferation and morphology.* Biotechnol Bioeng, 2012. 109(3): p. 695-707.

DePaola, N., et al., *Electrical impedance of cultured endothelium under fluid flow.* Ann Biomed Eng, 2001. 29(8): p. 648-56.

373309 Diffusion-Based Microfluidic PCR for "One-Pot" Analysis of Cells

Abstract:

Genetic analysis starting with cell samples often requires multi-step processing including cell lysis, DNA isolation/purification, and polymerase chain reaction (PCR) based assays. When conducted on a microfluidic platform, the compatibility among various steps often demands complicated procedure and complex device structure. Here we present a microfluidic device that permits "one-pot" strategy for multi-step PCR analysis starting from cells (i.e. using one reactor for successive reactions without separation and purification). Taking advantage of the diffusivity difference, we replace the smaller molecules in the reaction chamber by diffusion while retaining DNA molecules inside. This simple scheme effectively removes reagents from the previous step to avoid interference and thus permits multi-step processing in the same reaction chamber. Our approach shows high efficiency for PCR and potential for a wide range of genetic analysis including assays based on single cells.

So far there have been three strategies to alleviate the impact of cell pretreatment on PCR: 1. Institute some type of isolation step to remove lysis reagent and undesired intracellular molecules while preserving nucleic acids. The involved isolation step increase the complexity of the procedure and chip design. 2. Use alternative lysis methods, such as freeze-thaw or heating, that interfere with PCR to a less degree than surfactants. These methods are less efficient than surfactant-based lysis. 3. Use Direct PCR kit based on Phusion polymerase that is tolerant to surfactant-based lysis reagents, but this commonly-used fluorescence-based quantification is impossible with the Phusion polymerase system. To summarize, microfluidic strategies that permit simple operation and device design and are compatible with Taq polymerase are still in high demand.

In this work, we demonstrate a simple scheme for conducting microfluidic PCR starting from cells, taking advantage of the difference in the diffusivity between genomic DNA and various reagent/intracellular molecules. Our microfluidic device had a simple structure that included a reaction chamber connected with two loading chambers on both sides. The connections between the reaction chamber and the two loading chambers could be cut off by closing two-layer valves [7]. The lysis buffer and the PCR mix were introduced into the chamber by concentration-gradient-driven diffusion. During such diffusion, the new solution replaces the solution and molecules from the previous step without removing the slow-diffusing genomic DNA. The single chamber ("one-pot") design drastically minimizes the complexity of the microfluidic device. We envision that this may be a general approach for on-chip multi-step assays on genomic DNAs.

387315 High Throughput Microfluidic Separation of Tumor Initiating Cells (TICs) Using Contactless Dielectrophoresis

Abstract:

Tumor Initiating Cells (TICs), also known as Cancer Stem Cells, are a finite population of cancer cells possessing the unique ability to transplant a new tumor from an existing one and are putatively responsible for the metastatic properties of tumors. Contactless-dielectrophoresis (cDEP) is a non-invasive marker free technique able to characterize and enrich rare cells, such as TICs suspended in a low conductivity medium. cDEP separation is based on physical and electrical properties, and is independent of cell genotype. cDEP alleviates limitations of classical dielectrophoresis by using external electrodes placed in highly conductive channels which are separated from the sample by a thin insulating layer. We have developed the chip technology to make it more robust with increased throughput and increased yield for taking the enriched population off-chip.

When designing DEP separation devices the typical assumption is that the density of the cell suspension is low enough that cell-to-cell interactions can be neglected. That may be true for continuous flow through DEP devices however it is not true for DEP devices that trap cells on posts in areas with a high electric field. The force between induced dipoles (cells) contributes to the DEP force on a single cell level and pearl chaining is difficult to avoid. Therefore trapping depends not only on the trapping voltage of specific cells, but also on cell density (cells per ml). A possible solution that reduces this issue is to use a very low density of cells (~10^5 cells/ml), however throughput of such an approach is low. A better solution is to make the trapping area small enough that only one cell gets trapped while other cells are carried away by a sufficiently fast flow of media. A cDEP device was designed for trapping only one or two cells on a post at a time while maintaining high throughput. The diameter of the insulating pillars to make the electric field non-uniform were optimized to 20 μm which is close to the cell diameter. Due to the hydrodynamic drag force typically only up to two cells can trap on one pillar therefore pearl chaining and non-specific trapping is greatly reduced. The resulting high throughput separation device has about 17,000 insulating pillars 80 μm apart where cells get trapped with a main channel that is 12.3 mm long, 7.6 mm wide and 50 μm high. The developed protocol for separation is to pump cells suspended in DEP buffer at 2.5 μl/min and trap them at 300 Vrms, 100 kHz. When the insulating pillars are saturated with cells, syringes at the inlet are switched and pure DEP buffer is pumped into the device to wash non-trapped cells down the channel to the outlet for collection. The voltage is then gradually lowered, and cells are selectively released and separated cells are collected in tubes at the output. Separated cells are counted and analyzed. For assessing spheroid formation ability, a device with 4 parallel channels was used, providing enough separated cells for culturing in low-attachment dishes or alternatively cells can be analyzed by flow cytometer for expression of markers such as ALDH.

387349 On-Demand Control of the Limiting Current in Nano-Slot Devices By Varying the Diffusion Layer Length

Abstract:

The passage of an electric current through a permselective medium (membranes or nanochannel) under an applied electric field is characterized by the formation of ionic concentration gradients, which result in regions of depleted and enriched ionic concentration at opposite ends of the medium. In the low-voltage region, the current-voltage behavior is approximately Ohmic until the diffusion-limited current saturates when both ion concentrations are completely depleted at the surface. Here, we study the effect of controlling the limiting current, which roughly scales as the inverse of the diffusion length, in a microchannel-permselective medium system via varying of the diffusion layer length. The latter is varied by changing the distance of the reservoir or that of an on-demand convective mixing element from the nanoslot interface. We fabricated and tested nanoslot devices made of Nafion connecting two opposite microchambers made of PDMS. The device response was studied by visualizing concentration-polarization, colloid dynamics and measuring current-voltage. Enhancement of the limiting current by on-demand control of the diffusion length is of importance in on-chip electro-dialysis, desalination and preconcentration of analytes.

Coupling AC dielectrophoresis with DC ion concentration polarization in nanochannels for ultrafast biomarker enrichment

M. Tsegaye,[a] W. Varhue,[a] K. T. Liao,[b] C. F. Chou,[b] and N. S. Swami[a*]
[a] Department of Electrical & Computer Engineering
University of Virginia, Charlottesville, VA 22904, USA
[b] Institute of Physics
Academia Sinica, Taipei, Taiwan
*Corresponding Author's E-mail: nswami@virginia.edu

The charge selective transport characteristics at the entrance of nanochannels under conditions of ion concentration polarization (ICP) that arise due to the non-linearities in surface charge and field can cause sharp field gradients. In prior work these field gradients have been combined with electro-osmosis or electrophoresis, to cause exceptionally high degrees of biomarker enrichment [1]. However, since the trapped biomarkers are co-localized within a tightly confined region, the degree of selectivity is limited [2]. Dielectrophoresis (DEP) enables highly selective trapping of bio-particles based on the characteristic frequency response of the dielectric permittivity of the bio-particle versus that of the medium, and can be applied in principle towards frequency-selective biomarker enrichment [3]. However, DEP of nanoscale biomarkers is limited by the cube-fold drop in trapping force with particle radius and its disruption by electrothermal flow due to localized Joule heating within physiological media of high conductivity. While micro- and nanoscale constrictions can enhance the local field to enable frequency-selective trapping of ss-DNA [4] and proteins [5, 6] in physiological media, the highly localized nature of DEP behavior limits the spatial extent of biomarker depletion. Herein we explore the coupling of DEP behavior with concentration polarization for enhancing the spatial extent of biomarker depletion and thereby cause ultra-fast biomarker enrichment. We elucidate how the surface charge assymetry under DC fields within a nano-constriction device can be enhanced within physiological media through the application of AC fields to cause localized accumulation of counter-ions, thereby driving the ultra-fast enrichment of dielectrophoretically trapped biomarkers. This methodology has been applied towards enhancing the detection sensitivity of streptavidin, prostate specific antigen and neuropeptides [7].

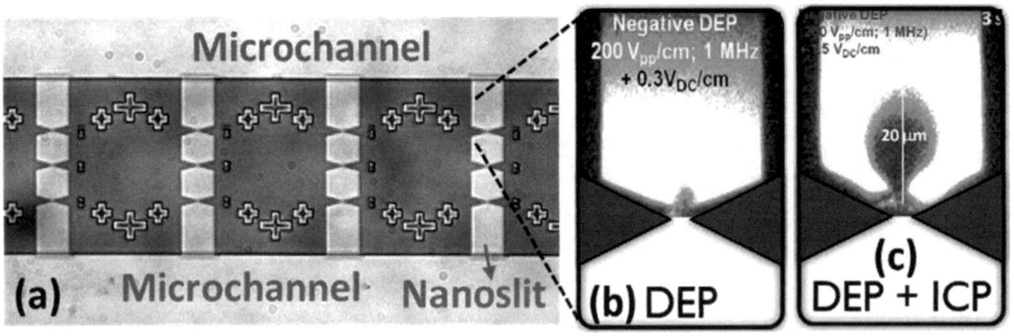

Figure 1: (a) Nano-slit channel with 50 nm constrictions; (b) DEP (200 V_{pp}/cm, 1 MHz) plus sub-critical DC field; (c) DEP with a critical DC field of 1.5 V/cm to initiate ion concentration polarization (ICP).

References

1. Y. C. Wang, A. L. Stevens, J. Han. *Anal. Chem.* **77**, 4293 (2005).
2. D. W. Inglis, E. W. Goldys,; N. P. Calander. *Angew. Chem. Int. Ed.* **50**, 7546 (2011).
3. R. Pethig. *Biomicrofluidics* **4** 022811 (2010).
4. N. Swami, C.-F. Chou, V. Ramamurthy, V. Chaurey. *Lab Chip* **9**, 3212 (2009)
5. K.T. Liao, M. Tsegaye, V. Chaurey, C.F. Chou, N. Swami. *Electrophoresis* **33**, 1958 (2012).
6. K.T. Liao, C.F. Chou, *J. Am. Chem. Soc.* **134**, 8742 (2012).
7. B. Sanghavi, W. Varhue, J. Chavez, C.F. Chou, N. S. Swami*; *Anal. Chem.* (2014), DOI: 10.1021/ac500155g.

369496 Tilted Post Arrays for Separating Long DNA

Abstract:

Recent simulations by Chen and Dorfman (Electrophoresis 35, 405-411, 2014) suggested that electrophoretic DNA separations in hexagonal arrays of posts would be improved by tilting the electric field with respect to the lattice vectors of the array. This improvement is attributed to the elimination of the channels between the posts that are present in normal post arrays. After the DNA collides with a post and unravels, it can travel in the channel for a substantial distance before it collides again. This effect is stronger at higher electric fields, typically greater than 20 V/cm, and results in the deterioration of separation quality at higher fields. In the tilted post array the DNA collides more often, but the posts are still in a regular array so the probability of multi-post collisions, which can significantly increase dispersion in the chip, is not increased. Overall, this results in a higher resolution separation or allows for the use of shorter post arrays.

We have constructed such an array where the electric field is applied at an angle equidistant between the two lattice vectors. The array leads to (i) baseline resolution of 20 kbp DNA and λ DNA (48.5 kbp) in a 4 mm channel and (ii) separations up to 50 V/cm, both of which are improvements over typical sparse post arrays. The predicted time required to reach a resolution of unity was approximately 5 mins, independent of electric field. The separations are more reproducible at higher fields.

371280 Electrophoretic Mobility of Nanoparticles Confined in Nanochannels

Abstract:

We investigate the mobility of a charged nanoparticle confined in a nanochannel driven by a weak electric field. Classic models for electrophoretic mobility are valid only in the linear regime of small particle zeta potential, and for an unbounded fluid domain. However, these models fail to predict electrophoretic mobility estimated from experiments using 50 nm diameter spherical particles in a 100 nm nanochannel.

We adopt asymptotically-expanded formulations and solve the fully-coupled equations on a 3D finite element domain to investigate such issues. Factors affecting particle mobility include electrolyte concentration, channel size, and zeta potentials on both the particle surface and channel walls. We find that nanochannel confinement combined with overlap of the double layers greatly affect the nanoparticle mobility. When the channel size is decreased from 2.5 um to 100 nm, the mobility is reduced by up to 20%. It also shows a possibility to use nanochannles to separate particles with the same zeta potential but different sizes. Finally, we propose a method to estimate corrected zeta potentials of the particle and walls from the observed mobility of a particle in both microchannels and nanochannels.

In addition, rotation is important when rod-shape particles are considered. We investigate the motion of a 2 nm diameter and 3.4 nm high rod in a 100 nm nanochannel in a 2D domain. Different initial positions and orientations are considered to observe the confinement effect. It shows the particle would be confined at the center of the channel and aligned with the applied electric field when double layers are thick (> 9 nm).

377148 Broken Symmetry in the Electrokinetic Flow Surrounding Asymmetric Colloidal Dimers

Abstract:

We report electric field induced propulsion of asymmetric colloidal dimers including both dielectric and metal-dielectric dimers. Under a perpendicularly applied AC electric field, the asymmetric dimers exhibit lateral motion on conductive substrates. Interestingly, the direction of the propulsion can be reversed by tuning both frequency and salt concentration. In addition, dielectric dimers can further assemble into chiral clusters, which behave like micro-propellers and rotate in opposite directions based on their handiness. However, all autonomous motions will disappear when the dimers stand up (i.e., align with the external field) at high frequencies. Neither spheres nor symmetric dimers show such propulsion under similar experimental conditions. All of the above observations can be explained by the symmetry-breaking in the electrohydrodynamic flow surrounding the dimers. By solving the standard electrokinetic model, we develop both a theoretical model and numerical simulations to investigate the detailed distribution of both electric and flow fields. Moreover, the delicate balance between the electrohydrodynamic and induced-charge electroosmotic flow, which have very different frequency dependence, could well explain the rich behavior that we have observed for the metal-dielectric dimers.

385602 Electrical Impedance Spectroscopy of Colloid-Nanoslot Interactions

Abstract:

Electrical Impedance Spectroscopy has been used extensively for the experimental study and characterization of a wide variety of electrochemical systems including ion-exchange membranes, micro-nanochannel devices, and colloidal suspensions. The interplay of electrokinetic transport mechanisms with colloid dynamics in the vicinity of micro-nanochannel interfaces is of both fundamental and applied interest. We present recent results concerning the characterization of colloid dynamics in the vicinity of a micro-nanochannel interface using electrical impedance spectroscopy. Frequency-dependent response of colloids in bulk solution and colloidal aggregation and accumulation at the nanochannel entrance under DC bias are compared and analyzed to determine the utility of impedance methods in the study of colloid dynamics and trapping.

386622 Electrokinetic Manipulation for Characterization and Capture of Circulating Tumor Cells

Abstract:

We describe dielectrophoretic manipulation of circulating tumor cells in hybrid DEP-immunocapture GEDI devices. Here, the difference in membrane capacitance between epithelial cells and hematological cells allows dielectrophoresis to be used to attract cancer cells to antibody-functionalized walls and reject contaminating leukocytes. Cell characterization and system design is informed by electrorotation studies of cancer cells, especially before, during, and after periods of vesicle shedding. Results are presented for prostate and pancreatic cancer cells.

New Applications of Electrophoretic Deposition (Solid State Lighting)

Jan B. Talbot

Chemical Engineering Program
Department of NanoEngineering
University of California, San Diego
La Jolla, CA 92093
jtalbot@ucsd.edu

UCSD Jacobs | School of Engineering

Electrophoretic deposition (EPD)

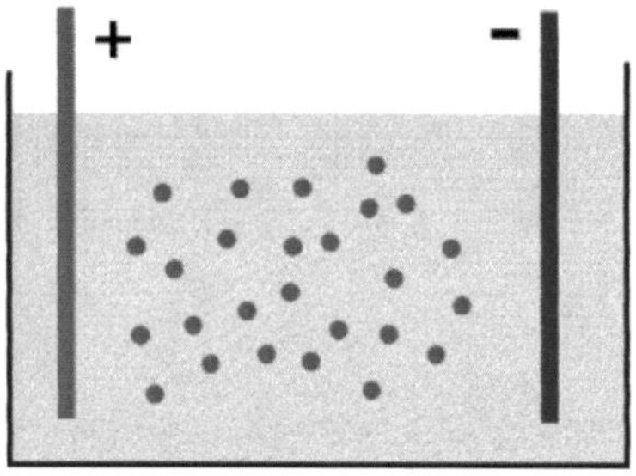

- Three major steps in EPD bath
  - 1) Charging of particles in suspension
  - 2) Transport of particles under an electric field
  - 3) Adherence of the particles to the substrate

Siracuse, Talbot, et al., J. Electrochem. Soc., 137 (1990)
Shane, Talbot, et al., J. Colloids and Interface Science , 165, 325-340 (1994).

Electrophoretic deposition (EPD)

- Powder film deposition process
- Short processing time
- Simple apparatus
- Easy modification
  - Flexibility of electrode shape
  - Control of thickness by the deposition time and applied voltage
- Deposit complex compounds and ceramics
  - Stoichiometry can be controlled by concentration in bath
- Applications to different materials
  - Silica, zeolites, high-Tc superconducting compounds, gas diffusion electrodes and sensors materials, oxide nanorods, carbon nanotube, piezoelectric materials, etc.

3

Applications of EPD

Carbonate particles on emitter wires for electron tubes

Electropainting of cars

Insulating glass for electronics packaging

Ceramic tapes, laminates, forms,

functionally-graded composites

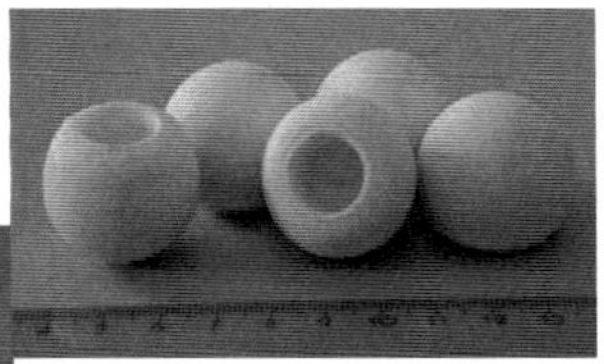

Solid oxide fuel cells

Nanostructured films

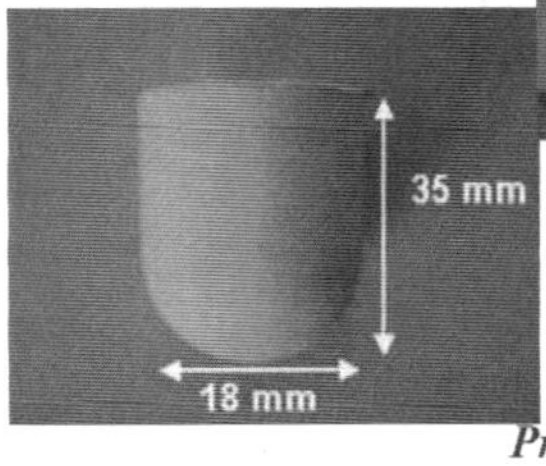

Prof. Van der Biest

http://www.mtm.kuleuven.ac.be/Research/C2/EPD.htm

UCSD Jacobs | School of Engineering

4

EPD Research at UCSD

PHOSPHORS (1990-present)
- ***Fundamentals of EPD***
- ***Adhesion***
- ***Full-Color (RGB) Deposition***
- ***Optical Characterization of EPD Films***
- ***SOLID STATE LIGHTING**

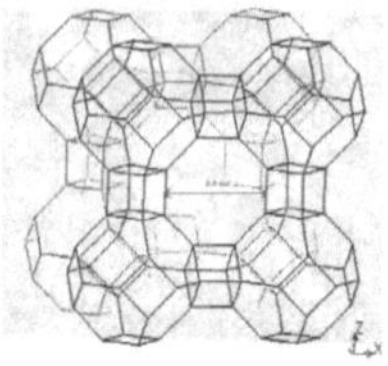

ZEOLITES
- ***Zeolite Modified Electrodes***
- ***Supported Membranes***

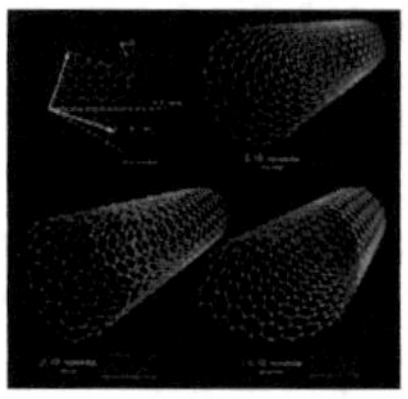

CARBON NANOTUBES

**ELECTROCATALYSTS*

UCSD Jacobs | School of Engineering

5

Lighting technology

Incandescent (~5% efficient)

Fluorescent (~20% efficient)

- Current lighting: poor conversion efficiency
- Highly efficient lighting is necessary
- Replacement with LEDs

Solid state lighting can reduce electricity needs for lighting by more than 60%

	Incandescent	Halogen	Compact fluorescent	LED
Energy efficiency	Very low	Low	High	Very High
Luminous efficacy (Lumen/Watt)	14	24	60-100	80-140
Lifetime (hours)	1000	2000-3000	6000-10000	50000
Heat	Yes++	Yes++	Yes	No
Mercury	No	No	Yes	No

DOE Solid-state lighting research and development: Multi-Year Program Plan

6

Generation of white light with LEDs

- **RGB LED**

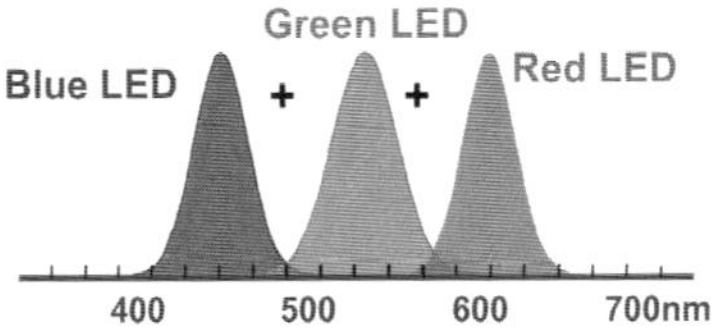

- Color variation with time and temperature
  - Low white color quality
- Severe current droop
  - Low efficiency at high current

- **Blue LED + Yellow phosphor**

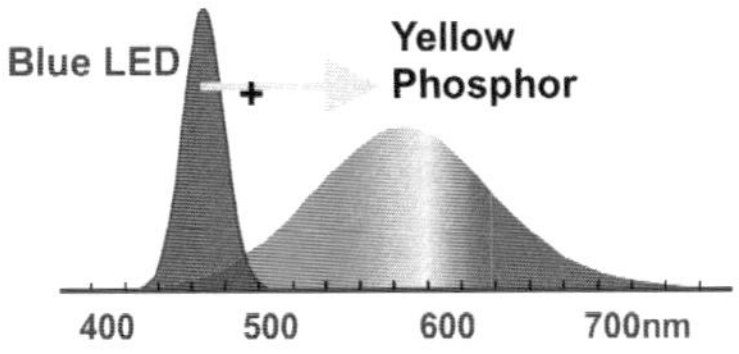

- Commercialized technology
- Low color rendering index (CRI)
  - Lack of red emission
- Current droop problem
- Color variation with input power

- **UV LED + RGB phosphor**

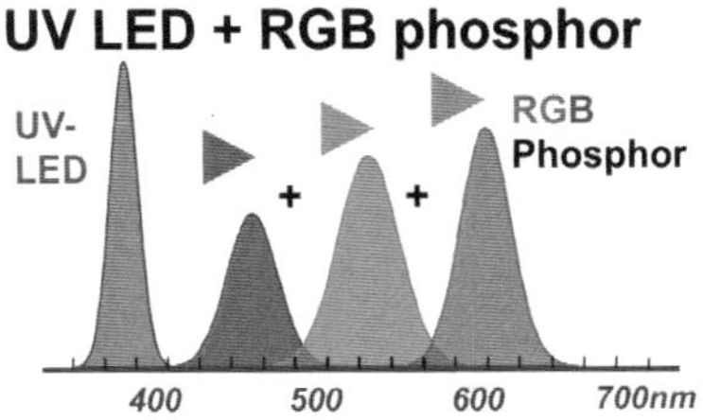

- Alternative method to produce white light
- Full color spectrum (high CRI)
- Alleviated current droop problem
- Less binning

7

CRI and CCT

Color Rendering Index (CRI)

- How accurately a light source renders color
- The CRI score is based on a set of 8 color samples illuminated by a reference source (black body)

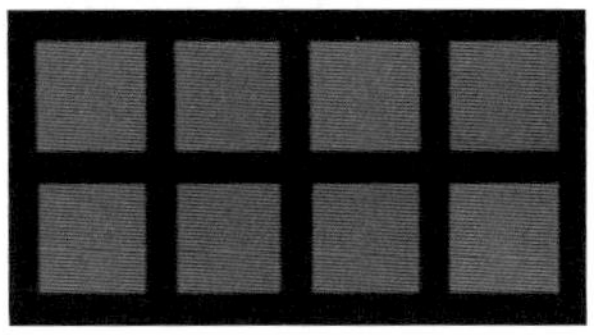

Correlated Color Temperature (CCT)

- A measure of warmth or coolness of the color of an artificial light source.
- CCT is based on a reference (black body) that is heated.
- White LED are classified as cool, neutral or warm
  - Warm white: 2700K – 3500K
  - Neutral white: 3500K – 4500K
  - Cool white: 4500K+

8

CIE diagram

1931 – The Commission International de l'Eclairage (CIE) defined a standard system for color representation.

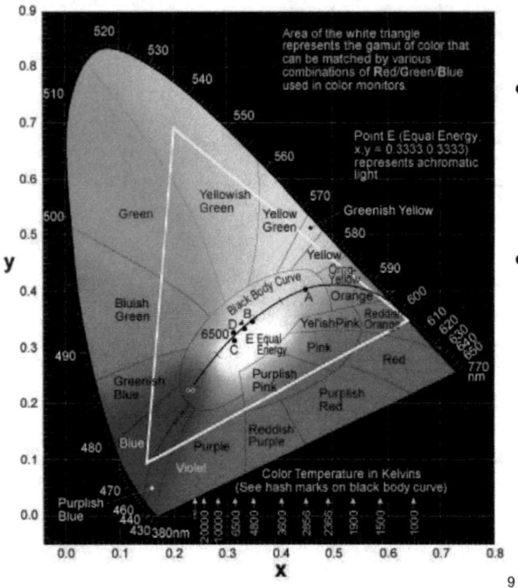

- Mixtures of red, green and blue
  - Any color in triangle
  - Optimal mixture yield white color
- On black body curve
  - Color temperature is determined

9 http://www.ledtronics.com/html/1931ChromaticityDiagram.htm

Objectives of our solid state lighting research project

- Electrophoretic deposition of phosphors
  - "Remote phosphor" configuration
- White light with optimal CRI (>80) and CCT (3000-4000K)
- Full conversion of UV light (thicker layer)
  - All LED light source emission → visible light
- Nano-sized phosphor deposition
  - Reduce scattering loss
  - Comparison between micron-sized and nano-sized phosphor

Work of Jake Choi, PhD (UCSD, now at Intel)

10

Remote phosphor configuration (EPD)

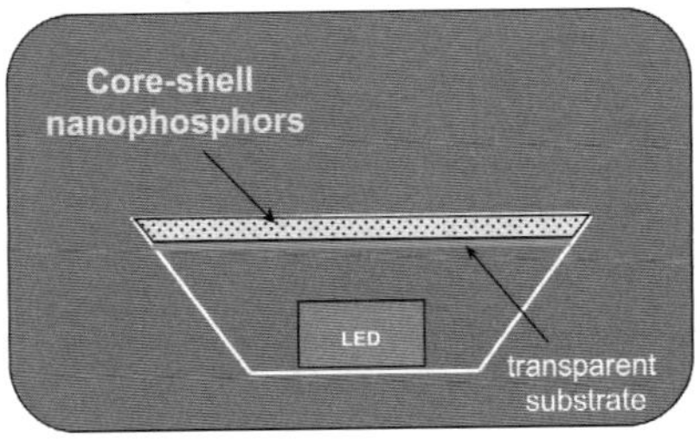

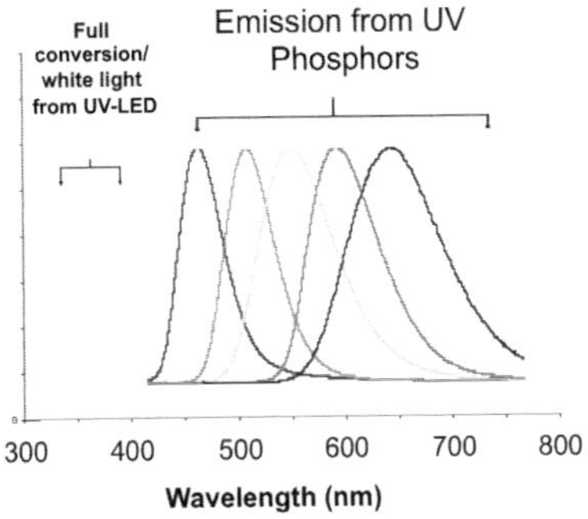

- Color blending flexibility
- Easy color quality control
  - CRI, CCT
- Improved extraction efficiency
- Thermal stability

➡ Improved light source design

11

Experimental set-up

1% phosphor concentration IPA solution
- 1 g/L phosphor + 2 vol.% glycerin + $Mg(NO_3)_2$ (10^{-5}-10^{-3} M)
- Applied voltage: 50-100 V
- Deposition time: 1-10 min.
- Substrate: ITO coated glass (1"×2"×1.1 mm, 70 ohms/sq)
- Electrode gap: 1-2 cm
- Zeta potential control by altering nitrate salt concentration (electric potential in the interfacial layer)

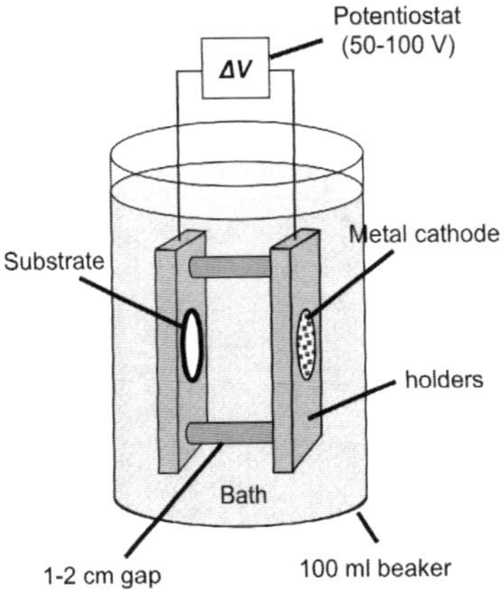

Chemical and electrochemical reactions

$2H_2O + 2e^- = 2OH^- + H_2$

$Mg(NO_3)^+ + 2OH^- = Mg(OH)_2 + NO_3^-$

12

Siracuse, Talbot, et al., J. Electrochem. Soc., 137 (1990)
Shane, Talbot, et al., J. Colloids and Interface Science , 165, 325-340 (1994).

Individual phosphor deposits

Photograph of individual phosphor EPD deposits

EPD bath

Deposit

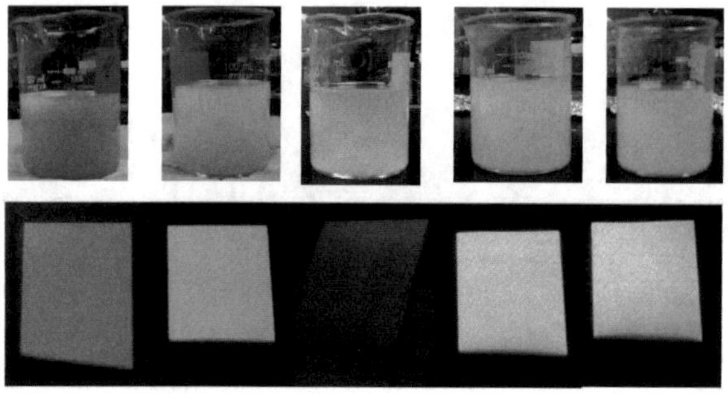

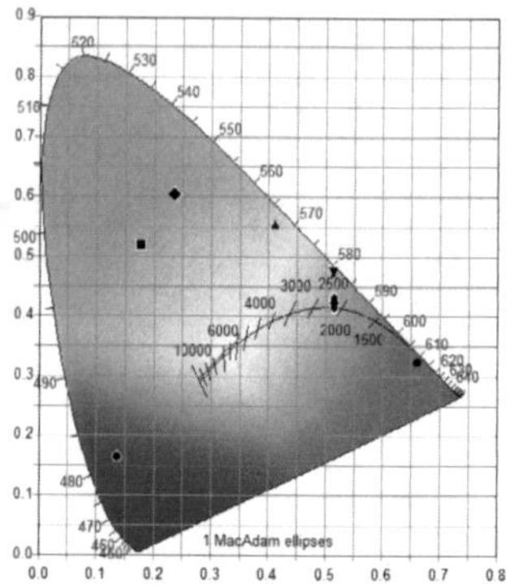

individual phosphor positions in CIE diagram

Eu-activated micron-sized powders
$Sr_{2-x}Ca_xSi_5N_8$ (red)
Ba_2SiO_4 (green),
$LiCaPO_4$ (blue)
$(Sr_{0.75}Ba_{0.25})_2SiO_4$ (yellow)
$(Sr_{0.5}Ba_{0.5})_3SiO_5$ (orange)

Optimal phosphor blends ➡ white light
Suggested by calculation ➡ determined by experimentally

13 J. I. Choi et al. Key Engineering Materials, 507, 149 (2012)

Deposition rate of single layer

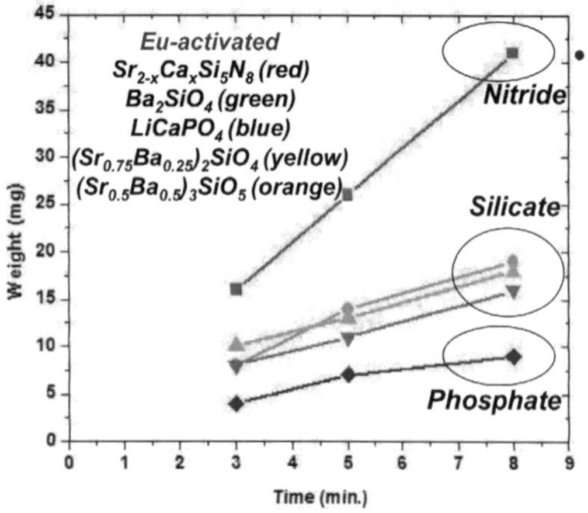

Deposit weight

- Nitride (red) phosphor
  - Thicker layer compared to others
  - Particles remained longer in the bath
  - Higher concentration at longer time
  - Reason is not understood
  - Further study is necessary
  (chemical environment and surface properties)

Due to settling: 8 min was maximum time to deposit particle (~10μm thickness)

J. I. Choi et al., ECS Journal of Solid State Science and Technology Vol. 2(7) R153 (2013)

14

Deposition types (sequential and blends)

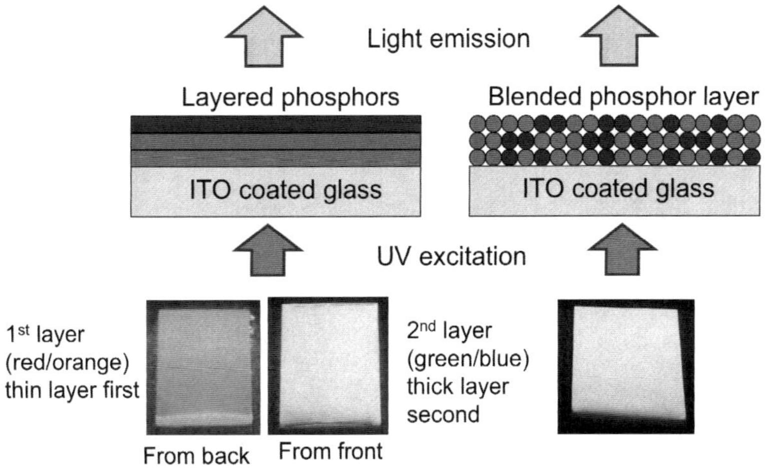

Multiple sequential deposition ***Single deposition***

Phosphor blend EPD is easier way to control the color coordinate

15

Phosphor blends for white light

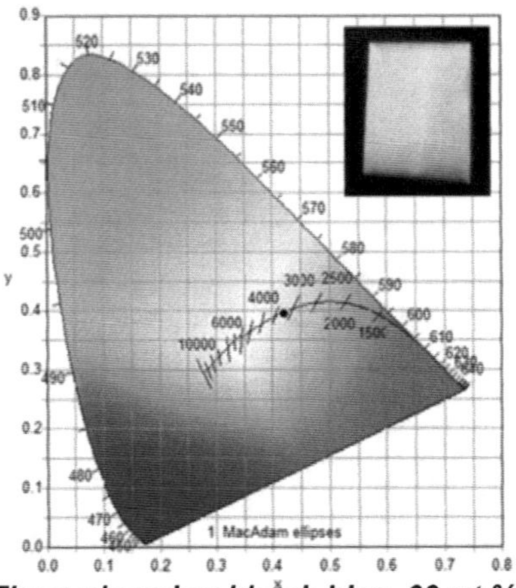

Three phosphor blend: blue, 63 wt.%, yellow, 15 wt.% and orange, 22 wt.%. CCT = 3202 K, CRI = 75.

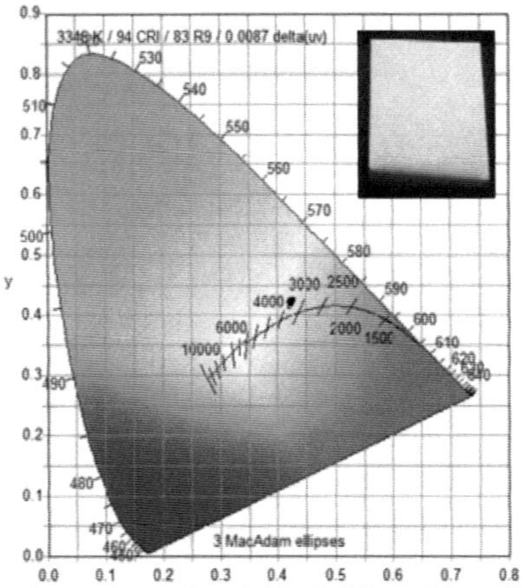

Four phosphor blend: blue, 70 wt.%, green, 13 wt.%, orange, 5 wt.% and red, 12 wt.%. CCT = 3346 K, CRI = 94.

Higher CRI

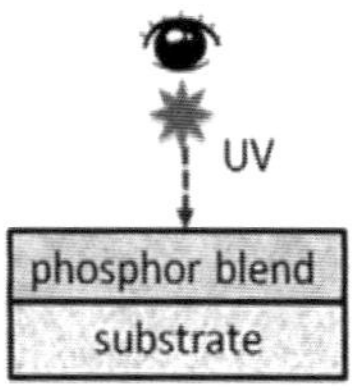

J. I. Choi et al., ECS Journal of Solid State Science and Technology Vol. 2(7) R153 (2013)

16

Sequential deposition

First layer (red/orange), second layer (green/blue)

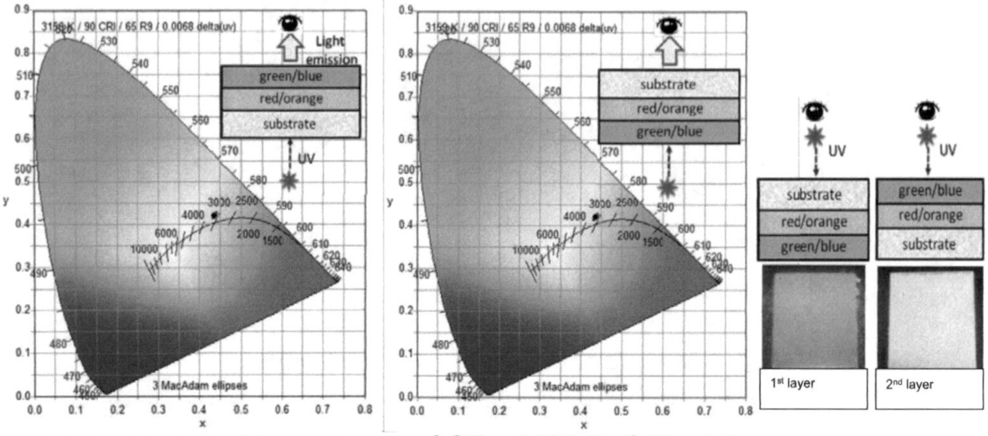

CCT = 3156 K, CRI = 90. **CCT = 3159 K, CRI = 90**

Layered films were flipped over: the same luminescence

J. I. Choi et al., ECS Journal of Solid State Science and Technology Vol. 2(7) R153 (2013)

17

EPD thickness for full conversion

- Thicker layer is necessary for full conversion
  - All LED light source emission → visible light
  - Thin layer: not fully converted, LED emission observed
  - Thick layer: full conversion (>~20 μm)

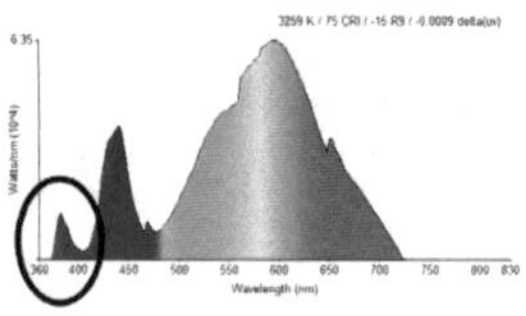

15 min EPD
UV light remains

Multiple EPD (5 min × 3 times)
- maximum thickness by EPD
- no UV light leak (full conversion)

18

Nanocrystalline phosphors

Particle size of commercial phosphors ~ 5-10 µm

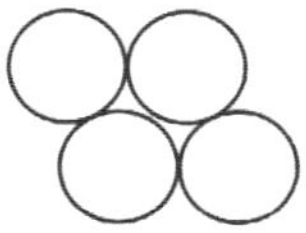

- ***Advantages***
  - *Easy to prepare*
  - *Known technology*
- ***Disadvantages***
  - *High scattering losses*
  - *Less scattering due to structural modification desirable - nanophosphor solution*

Nanocrystalline phosphors ~ 80-300 nm

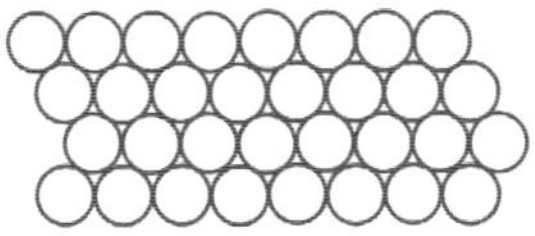

- ***Advantages***
  - *Reduction of scattering losses*
    - Potential to have higher λ_{ext} than micron-sized powders
- ***Disadvantages***
  - *Synthesis techniques not easy*
  - *Surface defects quench luminescence*
    - Core/shell methodologies

19

Nano-sized phosphor EPD from IPA bath

Ba_2SiO_4:Eu^{2+} green-emitting phosphor

micron ***nano*** ***micron*** ***nano***

Nano-sized phosphor EPD from IPA bath

:non-uniform and porous film (excessive hydrogen evolution by water)

- Increased surface area of nano-sized particle – more water adsorption

J. I. Choi et al., The Electrochemical Societyv 161, n 3, p D111-D117(2014)

20

Nano-sized phosphor EPD from amyl alcohol

IPA (Isopropyl alcohol)
- Water miscible (b.p. 83°C)
- Viscosity: 2.0 mPa·s
- Density: 0.786 g/cm^3)

Amyl alcohol (Isopentyl alcohol)
- Water is slightly soluble (b.p. 131°C)
- Viscosity: 3.7 mPa·s
- Density: 0.810 g/cm^3)

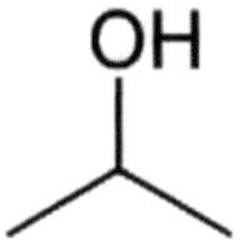

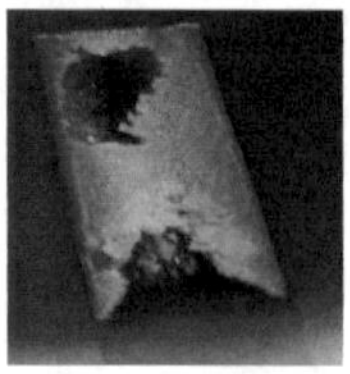

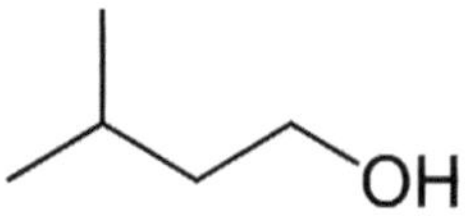

IPA bath (H_2 gas at interface)

Amyl alcohol bath

$2H_2O + 2e^- = 2OH^- + H_2$
$Mg(NO_3)^+ + 2OH^- = Mg(OH)_2 + NO_3^-$

New EPD bath is necessary: Amyl alcohol (water is slightly soluble)
- Careful control of the amount of water adsorbed

J. I. Choi et al., The Electrochemical Societyv 161, n 3, p D111-D117(2014)

21

New suspension medium (amyl alcohol)

Both nano- and micron-sized phosphors were successfully deposited

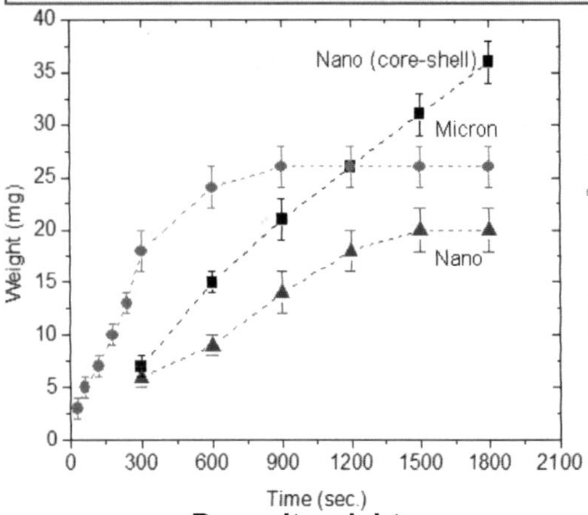

Deposit weight

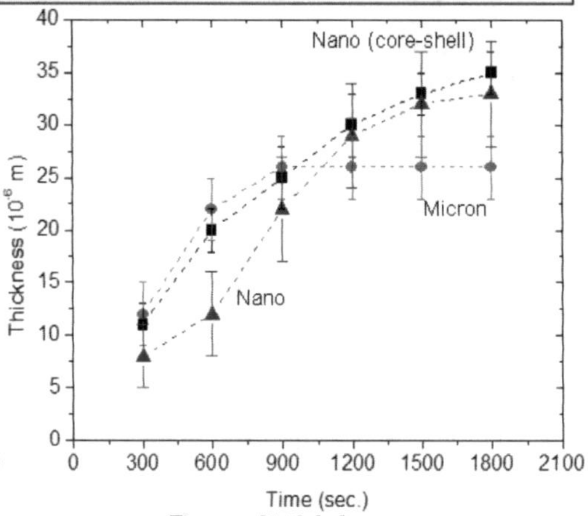

Deposit thickness

-Micron-sized: long time deposition (15 min, stable for a longer time) compared to IPA (8 min, fast settling effect) bath

J. I. Choi et al., The Electrochemical Society, v 161, n 3, p D111-D117(2014)

22

Cross-sectional SEM micrographs

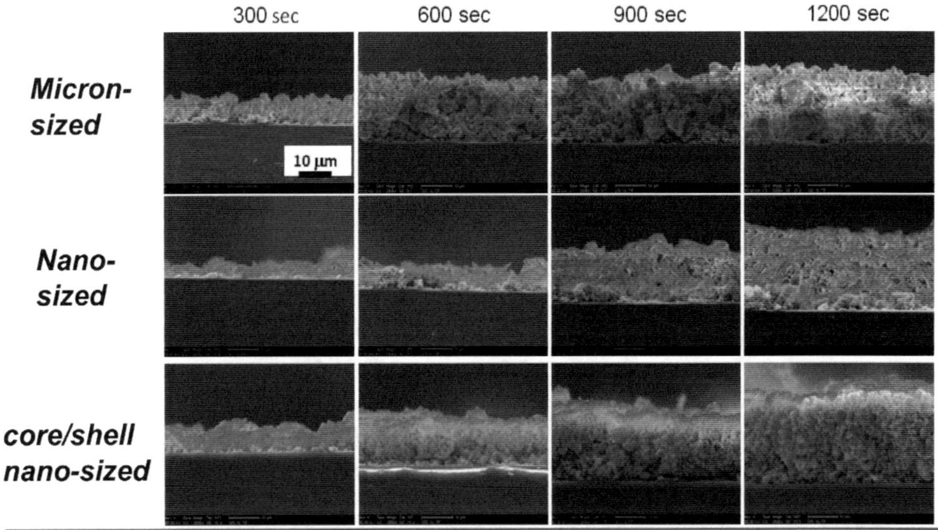

Film qualities are good for both micron-sized and nano-sized phosphor

J. I. Choi et al., The Electrochemical Society, v 161, n 3, p D111-D117(2014)

23

Modeling: two velocities (1)

Two velocities exist during EPD

1) Electrophoretic velocity

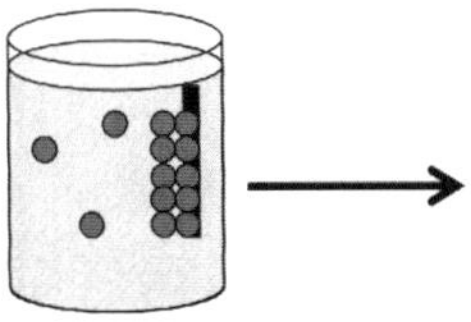

$$v_e = \mu E \qquad \mu = \left(\frac{\xi \varepsilon_r \varepsilon_o}{\eta}\right)$$

2) Settling velocity

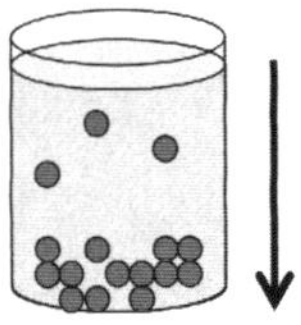

$$v_s = \frac{2}{9}\frac{(\rho_p - \rho_f)}{\eta} g r^2$$

v_e: electrophoretic velocity
μ : electrophoretic mobility
E : electric field
ζ : zeta potential
ε_r : relative permittivity
ε_0 : vacuum permittivity
η: viscosity
v_s: settling velocity
ρ_f : density of fluid
ρ_p :particle density
g : gravity
r : radius of particle

J. I. Choi et al., The Electrochemical Society, v 161, n 3, p D111-D117(2014)

24

Deposit weight prediction

Amyl alcohol bath

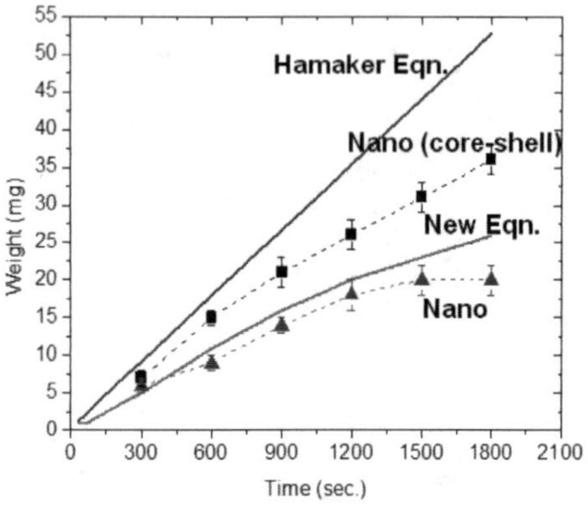

IPA bath

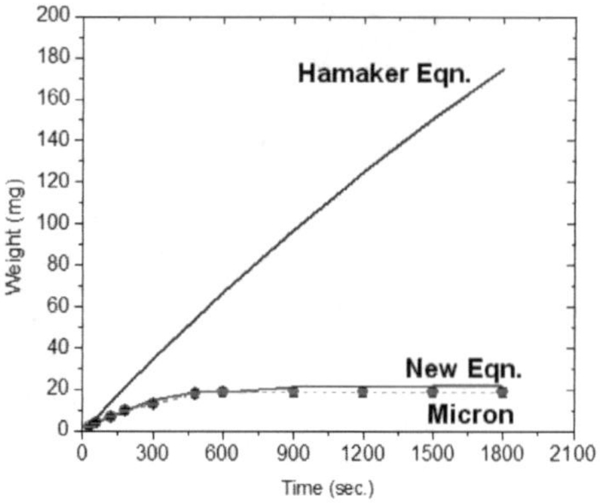

$m = C_S v_e A_S t$ ***: Hamaker equation (only valid for short times)***

New equation was developed

$$m(t) = \frac{C_{S,0} A_S v_e V}{v_s A}\left[1 - exp\left(-\frac{v_s A}{V} t\right)\right]$$

m: deposited weight
C_s: suspension concentration
v_e: electrophoretic velocity
A_s: deposition area, t: deposition time
$C_{s,0}$: initial concentration, V: bath volume
v_s: settling velocity
A: cross-section area of bath

J. I. Choi et al., The Electrochemical Society, v 161, n 3, p D111-D117(2014)

25

Conclusions

- Electrophoretic deposition is an effective way to deposit phosphors for solid state lighting
- Phosphor blends of three and four compositions produce white-emitting films with a CCT of ~3000K and a high CRI (>80)
- Thick layer (>~20 μm) is necessary to obtain full conversion
- For layered films of sequentially deposited phosphors with red/ orange and green/blue compositions
  - Order of phosphor deposition is not important
- For nano-sized phosphors EPD
  - Bath with low H_2O solubility required
- A new equation to predict the deposit mass was developed
  - Based on changing particle concentration in bath
  - Good agreement with experimental deposit weight

26

Acknowledgements

- Dr. Jake Choi (UCSD, now at Intel), Prof. J. McKittrick (UCSD)
- Dr. J.K. Han (UCSD, now at Brookhaven National Lab)
- Osram Sylvania solid state lighting research team
  - Dr. K. Mishra, M. Hannah, A. Piquette, and M. Anc.
- Dr. G. Hirata at UNAM
- E. Sluzky (Phosphor Consulting Service)
- This work is funded by the Department of Energy, Grant DE-EE-0002003

OSRAM SYLVANIA

U.S. DEPARTMENT OF ENERGY

27

386755 Bigger, Cheaper, Faster, More! DEP-Well Electrodes for Cell Electrophysiology

Abstract:

Dielectrophoresis has many advantages for cell electrophysiology. By fitting the frequency-dependent DEP behavior using the Clausius-Mossotti factor, it is possible to determine the conductivity and permittivity of the membrane and cytoplasm, and thereby understand the static electrophysiology of cells by relating the parameters to membrane morphology, membrane potential, cytoplasmic resistance and ion channel activity. It is simple to use, and cheap to implement; it can give insights into cell behavior that might require very expensive processes to reproduce - such as automated patch clamp, which typically samples only 40 cells but costs hundreds of thousands of dollars. However, whilst we are two years from the 50th anniversary of Herbert Pohl's demonstration of dielectrophoretic separation of live and dead yeast, the use of dielectrophoresis beyond "hard-core" dielectrophoreticists – those whose primary research line is dielectrophoresis, rather than those who just use it as a research tool– remains woefully low.

In order to find ways to bring DEP to the wider scientific community, we have tried to address the perceived shortcomings of the technique. We have done this by acting converse to academic instinct, by making devices bigger rather than miniaturizing, and simpler rather than more complex, but in such a way as to make them faster to operate and more easy to use. The resultant 3DEP DEP-well platform consists of multiple laminates of conductors (gold-plated copper) and insulators (polyamide) through which holes are drilled. Wells are typically 1-2mm deep and 0.5-1.5mm in diameter, containing about 10 electrodes. Since wells are drilled orthogonally to the laminates, many wells can be drilled side-by-side; routing of power lines can enable wells to receive different signals to their neighbors, or all can be wired together. A base can be added to form a closed well for particle analysis, or the well can be left open as a flow-through separator. The simplicity of the devices makes the consumable cost very low; devices can be made and the familiarity of the well-plate format makes it familiar to biologists; the cells resemble those found on 1536-well plates, and can be interfaced with standard fluid-handling robots.

When the field is applied, the cells move radially – either towards the center or the walls of the well – which can be tracked as a change in light intensity across the well. Parallel powering allows twenty parallel measurements to be taken; we have developed an integrated reader system containing 20 independent signal generators capable of reaching 50MHz, plus optics and camera. This allows a 20-point spectrum to be acquired in ten seconds, and the number of cells measured simultaneously (typically 20,000) means that data can be produced and fitted with Pearson coefficients greater than 0.99 with a single pass; data from multiple analyses can be averaged for very high accuracy data in a short period of time.

Without a base, the chips can be used for DEP separation. The size of the wells increases throughput, whilst the lack of high-aspect-ratio channels reducing pressure problems for chips placed in line with other microfluidic components, and parallelization allows very high throughputs to be achieved. However, the other great advantage of the device is speed; when used as a separator device, the devices can be built with hundreds of small-bore parallel wells, giving very high throughout (1 ml min^{-1} is not infeasible) cell separation with minimal cell loss due to the absence for a requirement of microfluidic tubing.

Applications have already been demonstrated in cancer diagnostics, cell biology, drug discovery and electrophysiology. For example, when used to analyze cells undergoing apoptosis using the 3DEP system to track proportions of healthy, affected and dead cells, it was found that IC50 measurements derived from DEP results correlated strongly with gold-standard MTT assays with significantly less preparation time and cost. Other applications include low-cost, point-of-care oral cancer diagnosis; it has been benchmarked against cardiomyocytes for understanding how the cytoplasmic resistance affects cardiac signal conduction; applications including erythrocyte chronobiology, cancer stem cell biology and stratified medicine, will be covered in detail elsewhere.

373747 Electrophoresis and Electroomosis in Planar Nanofluidic Channels

Abstract:

Electrokinetic based micro- and nanofluidic technologies provide revolutionary opportunities to separate, identify and analyze biomolecular species. In this talk, I will present results from joint theoretical and experimental studies of electroomosis and electrophoresis in planar nanofluidic channels. Key to fully harnessing the power of such systems is a thorough understanding of the influence of the electrokinetic surface properties. Therefore, I will also present results from investigations of electrokinetic surface properties in nano- and microfluidic channels fabricated with fused silica, insulated metals, and/or polymeric material. Specifically, I will describe techniques used to measure such electrokinetic properties within experimental micro- and nanofluidic systems, including the solution displacement method of current monitoring and electrokinetic injection techniques. Next, I will describe a models developed to find surface properties using this technique, and three separate experimental validations of this electrokinetic surface model, one using capillary filling, one with electrokinetic flow, and one using pressure driven flows. Finally, I will describe the latest results of electrophoretic separations in a nanofluidic channel with both small ions and short oligonucleotides, using combined electroomossis and pressure driven flow. These latest investigations uncover how to not only achieve infinite retention of ionic species using nanofluidics, but also how such channels can elucidate more information about the analyte of interest than conventional electrophoretic separation techniques.

386699 Field-Driven Dynamics of Metallo-Dielectric Particles and Particle Ensembles: From Programmed Assembly to Directed Motility and Actuation

Abstract:

Metallo-dielectric particles, such as Janus and patchy spheres, and selectively coated microcubes, acquire complex polarization pattern in electrical fields, which can lead to multidirectional dielectrophoretic interactions. In addition, such particles exhibit a variety of AC electrokinetic motility effects (also known as induced charge electrophoresis). The combination of dielectrophoresis and AC electrokinetic motility, controlled through the field frequency, strength and direction, opens a rich field of possibilities for engineered assembly and manipulation on the microscale. We will discuss the origins of the polarization and electrokinetic effects in systems of Janus and patchy particles and how they can be controlled and directed. One of the specific topics that has emerged from these studies is how complex directional polarization interactions (quadrupolar, hexapolar and multipolar) can be induced by external electric and magnetic fields and can be used to assemble new types of staggered chains, networks and crystals of unusual symmetry. We will examine how the particle assembly motifs and the structure of the small clusters formed can be analyzed via numerical calculations of the field intensity and energy in particle-medium polarization models. The assembly of permanent, yet reconfigurable, structures can be achieved with the additional use of magnetic fields and metal coatings with residual polarization. Overall, the field-driven assembly and dynamics of metallo-dielectric particles offers facile tools for making and actuation of numerous types of dynamic microscale structures, which may find applications in emerging fields such as soft robotics, microrobotics, and active and reconfigurable microfluidic devices. This potential will be illustrated in a preliminary way by a few examples of motile and self-folding particle assemblies.

Nonlinear Electrokinetic Effects on Particle Motion Near a Microchannel Constriction

Qianru Wang, Naga Neehar Dingari and Cullen R. Buie
Department of Mechanical Engineering, Massachusetts Institute of Technology

This work involves the first experimental observation of induced charge electrokinetic phenomena in insulator based dielectrophoresis devices. In this work, we experimentally and theoretically investigate the motion of particles in a microchannel with a 3D constriction[1,2] under alternating current (AC) electric fields. We also observe the rotation of *E. coli* under DC fields during 3DiDEP trapping.[1,2] Near the constriction, the particle or cell experiences two main forces, namely, the drag force due to background fluid flow and the combined effect of dielectrophoresis (DEP) and induced charge electrophoresis (ICEP) due to the nonuniform electric field, also known as dipolophoresis[3]. The background fluid flow near the constriction under AC fields is due to induced charge electro-osmosis (ICEO)[4] at the curved channel wall. Depending on the particle size, one of the three forces dominates over the others resulting in different particle motion patterns near the constriction.

We will show the motion of polystyrene beads of various sizes (200 nm, 1 μm, 6 μm) near the constriction. Particle motion near the constriction is reminiscent of ICEO flow near a curved surface.[4,5] The size of the vortices is larger for smaller beads (200 nm and 1 μm) indicating that the ICEO drag force dominates over the DEP force. For 6μm beads, the DEP force is large compared to the ICEO drag force, therefore the size of the vortex is smaller and the beads are repelled from the constriction due to negative DEP. In this work we neglect ICEP effects as polystyrene beads have very small surface conductivity, but this would be more significant for bacteria.

We substantiate our experimental observations with scaling arguments. The induced charge slip velocity in a curved microchannel scales as $u_s \sim \varepsilon_w E_0^2 L/\eta$, where L is the length scale of variation of channel dimensions. Hence the drag force acting on the particle scales as $F_{drag} \sim \eta a u_s = \varepsilon_w E_0^2 La$. Neglecting the induced charged effects on the particle, the dielectrophoretic force (F_{DEP}) acting on the particle scales as $F_{DEP} \sim \varepsilon_p a^3 \nabla E_0^2$. Near the constriction one can write $\nabla E_0^2 \sim E_0^2/L$. Taking the ratio of two forces, one obtains $F_{DEP}/F_{drag} \sim (\varepsilon_p/\varepsilon_w)(a/L)^2$. For very small particles we have, $a/L << 1$ and hence $F_{drag} >> F_{DEP}$ and particles move along ICEO flow paths, whereas larger particles are repelled away from the constriction due to negative DEP. We will provide design criteria for insulator based dielectrophoresis systems such that one might avoid or enhance these effects.

REFERENCES:

1. W. A. Braff, D. Willner, P. Hugenholtz, K. Rabaey, and C. R. Buie, *PLoS ONE*, 2013, **8**, e76751.
2. W. A. Braff, A. Pignier, and C.R. Buie, *Lab on a Chip*, 2012, **12**, 1327-31.
3. T. Miloh, *Phys. Fluids*, 2008, **20**, 063303.
4. T. M. Squires and M. Z. Bazant, *J. Fluid Mech.* 2004, **509**, 217–252.
5. G. Yossifon, I. Frankel, and T. Miloh, *Phys. Fluids* 2006, **18**, 117108.

388961 Fundamentals of Dielectrophoretic Particle Trapping in Arrays of Insulating Structures

Abstract:

This work is focused on insulator based DEP (iDEP), a dielectrophoretic mode that uses insulating structures between two external electrodes to create electric field gradients and generate dielectrophoretic forces on particles. This study includes experimental and mathematical modeling work. COMSOL Multiphysics was used to determine the distribution of the electric field and electric field gradient within an array of insulating posts. The distribution of these parameters was studied at each constriction between the insulating posts, in order to predict the net particle velocity over different electrokinetic and dielectrophoretic regimes. Experimentation was carried out with devices made from polydimethylsiloxane (PDMS) employing standard soft lithography techniques. All experiments were performed by applying DC electric potentials across the iDEP microchannels and observing particle response. The results clearly depict how particle trapping in iDEP devices occurs in the regions where the different forces acting on the particles are balanced. The findings from this study provide a deeper understanding of particle trapping in iDEP devices.

Dielectric Decrement Effects on Nonlinear Electrophoresis of Ideally Polarizable Particles

Jeffrey L. Moran,[1] Wai Hong Ronald Chan,[1] Bruno M. Figliuzzi,[2] Cullen R. Buie[1]

[1]*Massachusetts Institute of Technology, Cambridge, MA, USA*
[2]*Ecole des Mines de Paris, Paris, France*

We extend our previous work in numerically simulating the nonlinear electrophoresis of ideally polarizable particles to include the effects of a spatially non-uniform dielectric permittivity near the particle surface. Models for this dielectric decrement phenomenon have been developed by several authors, including Ben-Yaakov et al. (J Phys Condens Matter, 2009; J Chem Phys, 2011) and Hatlo et al. (EPL, 2012), and extended to electrophoretic phenomena by Zhao and Zhai (JFM, 2013). We extend this work to ideally polarizable particles and include the effects of surface conduction. By numerically solving for the coupled velocity field, electric potential, and ionic concentration distributions in the bulk solution surrounding the particle, we demonstrate that the dielectric decrement equations predict ionic saturation around the particle and thus physical implications that resemble those resulting from the steric model developed by Kilic et al. (Phys Rev E, 2007), albeit with differences that reflect the nonlinearity of the modified Poisson-Boltzmann equation. In addition, we develop a generalized condensed layer model that approximates both the steric and dielectric decrement models at high electric fields and low surface conduction to obtain more physical insights into the effects of these models. In particular, we demonstrate that the mobility in both models asymptotically scales as the square root of the electric field at high fields, once again recovering the result of Bazant et al. (Adv Colloid Interface Sci, 2009).

390790 Solution pH Changes in Non-Uniform AC Electric Fields Above the Electrode Charging Frequency

Abstract:

AC Faradaic reactions have been attributed as the mechanism inducing non-ideal phenomena such as flow reversal and cell deformation in microfluidics systems. Prior published work examined uniform electric fields below the electrode charging frequency (f_c), the frequency for electric double layer charging at the electrode. Spatially non-uniform AC electric fields are required for applications such as AC electroosmosis, AC electrothermal pumps, and dielectrophoresis. Many such applications utilize the AC frequency around or above f_c. In this work, fluorescein sodium salt, a pH sensitive dye, was used to detect $[H^+]$ in order to follow Faradaic reactions in aqueous solutions within non-uniform electric fields. Comparison experiments with a) parallel (uniform fields) electrodes and b) organic media were employed to test the electrode charging mechanism at $1.5f_c$. Time dependency analysis was processed and illustrated Faradaic reactions existence above the theoretically predicted electrode charging frequency. Further, spatial analysis revealed that the Faradaic reactions varied spatially due to the non-uniformity of the electric field and influenced local pH >50 μm away from the electrode edge. The non-uniform AC fields yielded pH gradients while uniform fields did not yield pH gradients, consistent with prior published data. Faradaic reaction frequency dependencies were examined from f_c to $3f_c$ at 11V peak-to-peak potential and voltage dependency was explored from 7 to $15V_{pp}$ potential at $1.5f_c$. Results imply that Faradaic reactions can still proceed within electrochemical systems in the absence of a well-established electric double layers. This work also illustrates that in microfluidics systems, medium variations need to be considered as a function of experiment time, initial medium conditions, electric signal potential, frequency, and spatial location in the fluid system.

383522 Electrically Induced Hydrodynamic Interactions in Capillary Electrophoresis of Polyelectrolytes

Abstract:

An electric field acting on a charged polymer and its counter-ion cloud can create a net velocity field at long ranges, though observing the phenomena requires that the polyelectolyte configuration be distorted from equilibrium. Hydrodynamic effects in electric fields are usually claimed to decay exponentially on a length scales proportional to the Debye screening length (κ^{-1}). However, the disturbance velocity in an electric field decays algebraically over the a distance r, with a leading order term of $(\kappa^2 r^3)^{-1}$. These electrically induced hydrodynamic interactions have been postulated to be responsible for the net transverse migration of DNA during transport through a capillary tube when both an electric field and pressure gradient are present [1].

Experiments have been conducted to test the hypothesis by measuring the migration of DNA when simultaneously applied an electric field and pressure gradient in parallel directions in a capillary. The shearing flow, created by the pressure gradient, acts to distort the DNA from its equilibrium configuration. Then, the hydrodynamic disturbances created by the electric field drive the migration. The extent of migration is measured as a function of electric field, flow strength, and Debye length, where the latter parameter is varied by changing the ionic concentration of the solution. The measured extent of migration is compared to the scaling predicted by the kinetic theory [1] and also to the results of Brownian dynamics simulations [2]. The results support the idea that velocity disturbances due to electrically induced hydrodynamic interactions exist and can have measurable influence on the dynamics of polyelectrolytes.

[1] J. E. Butler, O. B. Usta, R. Kekre and A. J. C. Ladd. Kinetic theory of a confined polymer driven by an external force and pressure-driven flow. Phys. Fluids, 19:113101, 2007.

[2] R. Kekre, J. E. Butler, and A. J. C. Ladd Role of hydrodynamic interactions in the migration of polyelectrolytes driven by a pressure gradient and an electric field. Phys. Rev. E, 82:050803(R), 2010

Examining frequency dispersion in non-linear electrokinetic flow using μPIV

Alicia Boymelgreen[1], Matan Zehavi[1], Gilad Yossifon[1]
[1]*Faculty of Mechanical Engineering, Micro- and Nanofluidics Laboratory, Technion – Israel Institute of Technology, Haifa 32000, Israel*

E-mail: aliciab@tx.technion.ac.il

Non-linear electrohydrodynamic flows, including induced-charge[1] (ICEO) and alternating-current[2] (ACEO) electroosmosis have received much attention amongst the micro/nanofluidic community over the past decade for their potential to produce net flow even under ac electric forcing, with suggested applications ranging from electroosmotic pumps[3] to externally controlled micromotors[4] and mixers. The advantages of ac electric forcing over its more commonly used dc counterpart include lower applied voltages and the suppression of Faradaic reactions and bubble generation which can block and degrade microfluidic systems.

The advent of such non-linear electrokinetic flows is strongly dictated by the frequency of the applied field. However, to date, there exists no unifying theory which can precisely predict both the strength and frequency dispersion of such electrokinetic flows, including the exact location of the maxima which is generally associated with the charge relaxation time of the system, $\tau = \lambda a/D$ (where a is the characteristic length of the system (for example, the distance between electrodes (ACEO) or radius of polarizable particle (ICEO)), λ the thickness of the Debye layer and D the diffusivity of the suspending medium). On either side of this maximum, ACEO flows are expected to decay as the electrodes become fully screened (DC limit) while at high frequencies (tens to hundreds of kHz) the flow tends to zero as the Debye layer at the electrode surface has insufficient time to charge. In the case of ICEO, maximum velocity is actually predicted in the DC limit (although in practical systems electrode screening and faradaic reactions usually prevent this from occurring) while again decaying beyond the charge relaxation time of the system.

Using μPIV and temperature sensitive dyes we demonstrate that these experimental tools can be used to assess the presence of additional physical phenomena such as dielectrophoresis, electrothermal flow and wall effects which can compete with induced-charge electrokinetic flow, influencing both the strength and frequency dispersion of the electrohydrodynamic flow. Within the context of this talk, we examine two very different geometries; asymmetric metallodielectric Janus particles and a weakly-polarizable dielectric channel corner. The results are of both fundamental and practical importance and may be used to further refine non-linear electrokinetic theory and optimize the flow parameters of electroosmotic pumps and the mobility of electrokinetically driven micromotors or carriers in lab-on-a-chip analysis systems.

1. Squires, T. M. & Bazant, M. Z. Induced-charge electro-osmosis. *J. Fluid Mech.* **509,** 217–252 (2004).
2. Ramos, A., Morgan, H., Green, N. G. & Castellanos, A. AC Electric-Field-Induced Fluid Flow in Microelectrodes. *J. Colloid Interface Sci.* **217,** 420–422 (1999).
3. Bazant, M. Z. & Squires, T. M. Induced-charge electrokinetic phenomena. *Curr. Opin. Colloid Interface Sci.* **15,** 203–213 (2010).
4. Boymelgreen, A., Yossifon, G., Park, S. & Miloh, T. Spinning Janus doublets driven in uniform ac electric fields. *Phys. Rev. E* **89,** 011003 (2014).

384427 Electromigration and Adsorption of Charge Carriers in Doped Nonpolar Liquids

Abstract:

Surfactants are added to nonpolar media to increase the electrical conductivity and control particle dispersion in a number of industrial applications. Steric stabilization of dispersions is typically assumed in nonpolar media; electrostatic effects are less well characterized. Electrochemical Impedance Spectroscopy (EIS) was used to analyze the electrical and physicochemical properties of dodecane containing up to 10 wt% of different surfactant 'dopants'. Commercial surfactants with the trade names OLOA 11000, Span 20, Span 80, Span 85, and Aerosol-OT (AOT) were studied. The nonpolar liquid solution was held between two parallel glass slides, coated with indium tin oxide (ITO) and held at a distance of 10 μm. Fitting the impedance of each solution to a 4-element equivalent circuit model provided the conductivity, dielectric constant, and Debye length. The conductivity of each surfactant solution was proportional to surfactant concentration. The Debye length of each surfactant solution was inversely proportional to the square root of the concentration. The number of charge carriers was calculated from the conductivity and Debye length measurements and found to be proportional to the number of surfactant molecules. Differences in the number of charge carriers per surfactant molecule between similar surfactants Span 20, Span 80, and Span 85 show that the structure of the nonpolar surfactant tail affects the number of charge carriers. The Nyquist impedance plots of AOT, Span 20, Span 80, and Span 85 solutions were qualitatively different from that of OLOA 11000. These differences suggest charge-carrier adsorption at the electrodes. More direct evidence of charge carrier adsorption was obtained using a Quartz Crystal Microbalance. Adsorption of surfactant and charge carriers may be an important mechanism for the acquisition of charge on solid surfaces in nonpolar media.

382751 Charging of Hydrophobic Polymer Particles By Basic Surfactants in a Nonpolar Liquid

Abstract:

Charging in nonpolar liquids is not common because of the large Born energy associated with ions in media of low dielectric permittivity. However, surfactants are known to reduce this energy cost and promote electric charging by increasing the effective ion size. Surfactants not only raise the conductivity of nonpolar bulk liquids but also cause surface charging of dispersed colloidal particles. Although particle charging in nonpolar liquids has been exploited in many practical applications such as electrophoretic image displays or electrostatic lithographic printers, the underlying mechanisms are still not understood. We discuss the charging of polymer particles in alkanes mediated by polyisobutylene succinimide-based surfactatns with systematic structural variations. The surface properties of the particle and surfactant polar head were characterized using the energy parameters of the van Oss-Chaudhury-Good (vOCG) theory of surface energy components. Based on the acid-base parameters of the interacting condensed phases and the electrophoretic mobility observed by phase analysis light scattering (PALS) for the charged particles and the particle-free micellar solutions, we propose that the charging of polymer surfaces may be achieved mainly by asymmetric adsorption of micellar ions. We discuss a hypothetical mechanism for the formation of micellar ion pairs with size asymmetry, and point out characteristic differences between the surfactants-mediated charging of hydrophobic polymer surfaces and that of hydrophilic surfaces in nonpolar liquids.

Conductive Hydrogel Membranes Produced by Electrophoretic Deposition at the Interface of Immiscible Liquids

Young Soo Joung,[1] Jeffrey L. Moran,[1] Robert Butler Ramirez,[2] and Cullen R. Buie[1]

[1]Department of Mechanical Engineering, Massachusetts Institute of Technology, Cambridge, MA, USA

[2]Department of Computer Science, Columbia University, the City of New York, New York, NY, USA

We present a novel method to fabricate composite nanomaterials using electrophoretic deposition (EPD) at the interface of two immiscible liquids. We demonstrate this method by creating composite hydrogel films. We denote this process as 'interfacial EPD'. During interfacial EPD, nanoparticles such as carbon nanotubes (CNTs) migrate to the oil/water interface, where cross-linking of polymers is induced to form composite hydrogel membranes. The key aspect of this method is that polymerization occurs away from a solid substrate while surrounded by both polar and nonpolar media, allowing for the integration of CNTs or other nanoparticles into the hydrogel. Advantages of material assembly at the immiscible liquid interface include scalability and the ability to create free standing thin films while avoiding difficulties associated with removal from a solid substrate. Properties of the composite hydrogel films are controlled by the deposition and polymerization time, allowing for potential mass production without the need for complex machinery. This fabrication method is cost-effective and scalable for composite hydrogels with tunable electrical, mechanical, and biological properties. Potential applications include fabrication of doped hydrogels for drug delivery and conductive hydrogels for biological sensing.

390501 Microfluidic Mixing of Nonpolar Liquids By Contact Charge Electrophoresis

Abstract:

The manipulation of colloidal scale particles through electrophoresis is a powerful tool for driving basic unit operations such as separation or mixing processes, particularly in microfluidic systems. Contact charge electrophoresis (CCEP) denotes a specific category of electrophoresis where the particle actuated by an electric field acquires charge when in contact with an electrode. This mechanism allows a particle to achieve high frequency oscillations (~100Hz) between parallel electrodes, and can be coupled to simple geometries such as ramps to direct particles transverse to the electric field. Here we demonstrate a simple microfluidic mixer which uses CCEP to achieve rapid effective stirring of two dielectric liquids by stretching and folding the flowing streams into one another. Interestingly, the simple reciprocal motion of CCEP alone was largely ineffective at mixing two laminar flowing streams due to the kinematic reversibility of low Reynolds number flows. The incorporation of PDMS ramps which guide the particle in an orbital trajectory is critical to achieve rapid mixing. We discuss the optimal operating conditions for CCEP-based mixing, and investigate CCEP in 15 common organic solvents to determine which liquids can be effectively mixed using this approach. As CCEP only requires a low power DC voltage, this mixer is ideally suited for applications in portable microfluidic devices.

385055 Contact Charge Electrophoresis for Powering Micro- and Nanotechnology

Abstract:

Electric fields have long been used to manipulate colloidal matter (*i.e.*, micro- and nanoscale particles, droplets, macromolecules) and to control its organization. This talk will describe a relatively new form of electric particle actuation called **c**ontact **c**harge **e**lectro**p**horesis (CCEP), in which a conductive object is first charged by contact with an electrode surface and then actuated by an externally applied electric field. Unlike common forms of electric actuation (*e.g.*, dielectrophoresis), CCEP allows for rapid, sustained motion driven by low power DC voltages. As a result, this mechanism is well suited for powering the active components of mobile microfluidic technologies and may provide a basis for efficient, chemically-powered colloidal machines. This talk will describe our recent efforts to understand the fundamental physics underlying CCEP and its application within microfluidic systems.

365095 Continuous Label-Free Particle Separation Via Wall-Induced Lift in Electrophoresis

Abstract:

The efficient separation of discrete particle (both synthetic and biological) species is a topic of interest in numerous research fields for its practical application to problems encountered in both academia and industry. We demonstrate in this work a novel continuous-flow label-free method for electrokinetic separation of particles based on size and/or surface charge. This method utilizes the wall-induced non-inertial lift force to deflect a sheath flow-focused particle mixture to property-dependent positions in a straight microchannel. A numerical model is also developed to understand this separation and to study the parametric effects on it. The numerical predictions are found to agree reasonably with the experimental observations.

AN ELECTROPHYSIOLOGICAL STUDY OF CHEMOTHERAPEUTIC AGENTS ON CANCEROUS CELLS USING DIELECTROPHORESIS (DEP)

Sina Mahabadi, Fatima H. Labeed and Michael P. Hughes

Centre for Biomedical Engineering, University of Surrey, Guildford, Surrey GU2 7XH, United Kingdom

Email s.mahabadi@surrey.ac.uk

Sin the first demonstration of the separation of live and dead yeast in 1966, dielectrophoresis (DEP) has been widely employed for examining the biophysical basis of the differnces between different cells. Electrophysiological properties such as the conductivity and permittivity of the membrane and cytoplasm of the cells can be obtained through using DEP through the fitting of a single-shell model to spectra correlating to the frequency-dependent change in the Clausius-Mossotti factor.

Head and neck cancer is the 8th most common worldwide, and cisplatin and Iressa (Gifitinib) are two of the most widely used anti-cancer treatments for this disease. In this work, the effect of Iressa and cisplatin on electrophysiological properties of HN5 (head and neck carcinoma) was investigated. The cells were grown in standard cell culture conditions, and they were tested before incubation with the drugs, and after 22 hours of drug treatment. DEP measurements were performed using the 3DEP electrode microwell system developed at the University of Surrey.

The results indicated a marked change in cell membrane capacitance of HN5 cells when cisplatin and Iressa were combined for the treatment. In contrast, the cells showed a much smaller decrease in membrane capacitance when cells had undergone a single drug treatment.

Our study demonstrated that the combination of the two drugs remarkably affect the electrophysiological properties of head and neck cancer cells. This further demonstrates that DEP is a useful technique to assess different approaches of anti-cancer drug treatments, and may have potential applications in stratified and personalized medicine for the identification of the most appropriate combination of drugs for an individual.

Insulator Based Dielectrophoresis to Correlate Cell Polarizability and Electrocompetency

Zhifei Ge, Jeffrey Moran, Paulo Garcia, Cullen Buie

Department of Mechanical Engineering, Massachusetts Institute of Technology

Escherichia coli have been widely used as the chassis for bacterial transformation in synthetic biology [1]. However, *E. coli* has its limitations. Promoters from some species, such as *mycobacteria*, do not all function in *E. coli* [2]. Moreover, all posttranslational protein modifications do not take place in *E. coli.* [3] To fully tap into potential and explore possibilities of synthetic biology, more chassis for transformation have to be built. The key to building a chassis for synthetic biology is to have optimal protocols to introduce genes into cells.

To expand the pool of bacterial species that can be reversibly electroporated with efficiency [4], we propose a field-assisted method to search for optimal protocols of reversible electroporation. In our method, we hypothesize that polarizability of the cell under electric field correlates with the cell's electrocompetency, and higher polarizability represents higher electrocompetency. With this hypothesis, we can predict electrocompetency by measuring polarizability. The key innovation in our method is that we predict transformation efficiency before the cell is electroporated, while other existing method requires that the cell be electroporated and the electroporated cell be cultured.

Both dielectrophoresis and electroporation rely upon redistribution of charges surrounding and inside the cellular membrane. Due to the common link of cell membrane polarization, we hypothesize that membrane polarizability and electrocompetency are correlated, with higher polarizability corresponding to higher electrocompetency. In this work, we will present a framework to validate this correlation and initial experimental results to strengthen this hypothesis.

We use 3D insulator based dielectrophoresis [5] to quantify polarizability. Poly(methyl methacrylate) (PMMA) is used to build the 3D insulator-based dielectrophoresis chip, in which a 50 um X 50 um constriction channel is designed to connect the 500 um X 500 um channels. Near the constriction area, high electric field gradient is generated. Under DC electric conditions without external pressure driven flow, forces on the cell will include electrophoretic force, electroosmotic force (dragging force), dielectrophoretic force, lift force, and gravity. Neglecting lift force and gravity, balance of the other three forces can trap cells in regions near the constriction area. Average fluorescence is used to measure average polarizability in a cell population. When a certain number of fluorescent cells are seeded into the microchannel, a higher number of cells will be trapped when the average polarizability is higher, which is quantified by the fluorescent intensity near the constriction area.

Our initial experimental results show that *E. coli*'s polarizability changes during growth correlates with electrocompetency. Our experimental data show that exponential phase *E. coli* have higher polarizability, compared with stationary phase. Note that it is well known that exponential phase *E. coli* are more electrocompetent than stationary phase *E. coli*. Our preliminary experimental results strengthen our hypothesized correlation between polarizability and electrocompetency.

References

[1] E. Andrianantoandro, S. Basu, D. K. Karig, and R. Weiss, "Synthetic biology: new engineering rules for an emerging discipline," Molecular Systems Biology, vol. 2, May 2006.

[2] L. Kremer, A. Baulard, J. Estaquier, A. Capron, and C. Locht, "Analysis of the Mycobacterium tuberculosis 85A antigen promoter region.," Journal of bacteriology, vol. 177, no. 3, pp. 642–653, 1995.

[3] M. Tokunaga, H. Tokunaga, and H. C. Wu, "Post-translational modification and processing of Escherichia coli prolipoprotein in vitro," Proceedings of the National Academy of Sciences, vol. 79, no. 7, pp. 2255–2259, 1982.

[4] M. Keller and K. Zengler, "Tapping into microbial diversity," Nat. Rev. Microbiol., vol. 2, no. 2, pp. 141–150, Feb. 2004.

[5] E. B. Cummings and A. K. Singh, "Dielectrophoresis in Microchips Containing Arrays of Insulating Posts: Theoretical and Experimental Results," Analytical Chemistry, vol. 75, no. 18, pp. 4724–4731, Sep. 2003.

387665 Micronanotip Injection Electroporation

Abstract:

DNA or RNA-based non-viral therapy methods have been developed to serve as favorable alternative in cell therapy to their virus-mediated counterparts while still suffer some unsatisfied performance (e.g., low cellular uptake, high cytotoxicity, and/or low efficiency). We envision that the introduction of micro/nanostructures on electrodes could largely advance electroporation-mediated DNA and RNA delivery to mammalian cells. To test this hypothesis, we develop both micro-nanotip injection electroporation platforms and evaluate their contributions to the delivery efficiency of DNA and RNA probes with both anchor cells and suspension cells. The micro/nanostructures offer mild electroporation conditions through highly focused pulse strength and localized poration to the cell membrane. Single or multiple exogenous genes can be programmed to the same cells in a dosage-controlled manner. The well-defined patterns ensure the same treatment to a large number of cells of various sizes. This may bridge two important, but long separated electroporation fields: single cell electroporation and cell electroporation of a large population (bulk electroporation). In this way, it could not only greatly improve gene delivery efficiency and cell viability, but also facilitate mechanism discovery on cellular uptake dynamics and simplify the tedious, cell-specific protocol searching process. Its success will benefit the whole life science and biomedical community where a safe and effective non-viral gene delivery approach is needed on a daily basis.

373255 Microfluidic Electroporation for Delivery of Cell-Penetrating Peptide Conjugates of Peptide Nucleic Acids (PNA) for Antisense Inhibition of Intracellular Bacteria

Abstract:

Cell penetrating peptides (CPPs) have been used for a myriad of cellular delivery applications and were recently explored for delivery of antisense agents such as peptide nucleic acids (PNAs) for bacterial inhibition. PNAs, gene-specific artificial oligonucleotide analogs with peptide backbone, are designed to silence selected gene and inhibit bacteria. Although these molecular systems (i.e. CPP-PNAs) have shown ability to inhibit growth of bacterial cultures *in vitro*, they show limited effectiveness in killing encapsulated intracellular bacteria in mammalian cells such as macrophages, presumably due to difficulty involved in the endosomal escape of the reagents.

In this report, we show that microfluidic electroporation delivery dramatically increases the bioavailability of CPP-PNAs to kill *Salmonella enterica* serovar Typhimurium LT2 inside macrophages which is typical in animals and humans. Electroporation (i.e. the exposure of cells to an external electric field with a higher-than-threshold intensity for a short duration) breaches cell membranes by creating nanoscale pores and destabilizing the membrane structures. Electroporation delivers the molecules without involving endocytosis and greatly increases the antisense effect. The decrease in the average number of *Salmonella* per macrophage under a 1200 V/cm and 5 ms pulse was a factor of 14 higher than that without electroporation (in an experiment with a multiplicity of infection of 10:1). Our results suggest that electroporation is an effective approach for a wide range of applications involving CPP-based delivery. The microfluidic format will allow convenient functional screening and testing of PNA-based reagents for antisense applications.

Electroporation results from exposure of cells to electric fields of sufficient strength to disrupt the plasma membrane of the microorganisms [1]. The local trans-membrane voltage (TMV) significantly increases during exposure to external electric fields. When the local TMV exceeds a critical threshold value, pores are created on the membrane of the cells, allowing for transport of ions and macromolecules across the membrane [2]. This membrane disruption results in a non-linear increase in the electrical conductivity of the membrane that spans several orders of magnitude. Associated with the increase of the electrical conductivity of the membrane is the increase of permeability of foreign molecules, including foreign DNA, that may translocate through the pores and into the cell. There is vast empirical evidence in the literature establishing protocols to increase the electrocompetency of cells in a process that is time consuming and without real-time feedback. This is performed in practice because higher electrocompetency results in higher efficiency in the genetic electrotransformation. However, there are currently no protocols for measuring and improving the electrocompetency of bacterial species without going through the empirical electroporation process.

We have developed a rapid microfluidic assay to visualize and characterize the electric field required for successful electroporation of microorganisms, one of the parameters for determining electrocompetency. Other flow-through electroporation methods have been developed by Geng *et al.* [3] for delivery of genes into cells and by Wang *et al.* for electrical lysis of bacteria [4]. Our diagnostic assay differs from that of the group of Lu *et al.* [5] in that our aim is to rapidly characterize the critical electric field strength required for inducing genetic transformation for currently intractable cells whose transformation potential remains unexplored. The assay is based on a microfluidic channel with a configuration that contains three distinct regions. The first region depicts the inlet and outlet reservoirs that also connect to the pulse generator via stainless steel electrodes. The second region is an area that results in a constant low electric field. Finally, the active region of electroporation spans 2-4 mm and has a geometry that enables the generation of high electric fields (~ 5-15 kV/cm) required for successful transformation of bacteria. This converging geometry was optimized to result in a semi-linear and continuous spectrum of electric field strengths.

The vision of this rapid assay is to use a single chip to provide microbiologists with the experimental tools to quantify the electric field that will provide the greatest chance at successful genetic transformation. Our work differs from the traditional trial and error approach employed in the research community in that we develop an assay for systematically sampling a continuum of electric fields that can be used to probe any type of bacteria or potentially perform on-chip electrotransformation. Ultimately this work provides a reliable platform to rapidly determine electroporation conditions for bacterial strains for which traditional field assisted genetic transformation has been unsuccessful. Results of this study will broaden the scope of bacteria available for applications in synthetic biology and genetic engineering.

1. Weaver, J.C., *Electroporation: a general phenomenon for manipulating cells and tissues.* J Cell Biochem, 1993. **51**(4): p. 426-35.
2. Weaver, J.C. and Y.A. Chizmadzhev, *Theory of electroporation: a review.* Bioelectrochem. Bioenerg., 1996. **41**: p. 135-60.
3. Geng, T., et al., *Transfection of cells using flow-through electroporation based on constant voltage.* Nat. Protocols, 2011. **6**(8): p. 1192–1208.
4. Wang, H., A. Bhunia, and C. Lu, *A microfluidic flow-through device for high throughput electrical lysis of bacterial cells based on continuous dc voltage.* Biosensors and Bioelectronics, 2006. **22**(5): p. 582-588
5. Geng, T. and C. Lu, *Microfluidic electroporation for cellular analysis and delivery.* Lab on a Chip, 2013. **13**(19): p. 3803-3821.

Coordination of many of the body's behavioral, metabolic, and physiological processes are controlled around the 24-hour clock through endogenous circadian rhythms which can be synchronised (entrained) by external stimuli, which allows for gradual readjustments with the natural changes in photoperiod, and for abrupt adjustments such as following transmeridian travel. Disruption of these circadian rhythms can lead to negative health outcomes related to areas such as metabolism, cardiovascular health, and mental health. Understanding these rhythms can help determine optimal timing of physical and mental activities as well as drug delivery.

There is ample evidence that the main drivers of the process are specific clock genes which drive rhythmic expression over a near-24-hour cycle. Disruption of these genes in mammals leads to disruptions in circadian rhythms. This paradigm was recently challenged by the finding that following temperature entrainment, circadian rhythms in redox status are present in human red blood cells (RBCs) [1]. RBCs have no nucleus and thus no transcription. What other mechanisms are controlling rhythms present in RBCs, and how are they generated? Although the finding of transcription-independent circadian rhythmicity has attracted considerable interest, it has not been corroborated in the published literature. Here, we used dielectrophoresis (DEP), an electrostatic technique used to examine cellular electrophysiology, to human erythrocytes cultured under the same paradigm, to demonstrate rhythmic perturbations in the electrophysiological properties of human RBCs.

DEP well chips [2], which consist of an array of 20 wells, have been designed along with the 3DEP reader to simultaneously measure cellular response at 20 different frequencies. The reader collects images of the wells every 0.5 s and tracks changes in light due to cell movement. The changes in light intensity are related to DEP force, and after each experiment a DEP "spectrum" is obtained as a function of frequency [2]. The spectra are then fit to a single-shell dielectric model to obtain electrophysiological properties of the cells.

Four male donors, free of self-reported health problems including sleep disorders, provided 10-ml blood samples in heparin/EDTA tubes. RBCs were purified and entrained in two consecutive cycles of 12 h at 37°C and 12 h at 32°C, after which cells were left to free-run at 37°C for the duration of the incubation [1].

Starting at Zeitgeber time 0h, 500µl samples were removed at four hourly intervals for DEP analysis over a total period of 48 h. Samples were washed in isoosmotic DEP medium [4] adjusted to a conductivity of 0.043 S/m using phosphate-buffered saline (PBS), and resuspended at a final cell concentration of 10^6 cells/ml. Cell suspensions were pipetted in DEP well chips (DEPtech, Uckfield, East Sussex, UK) for analysis. Each time point was repeated at least 3 times.

Raw data were fitted to a single-shell model [3] ($R<0.9$ rejected), through which values for the effective membrane conductance (describing ion transport across the membrane) and capacitance (reflective of membrane morphology), and cytoplasmic conductivity (numerating ionic strength) were obtained [2]. One-way ANOVA was carried out in MATLAB to examine the effect of time on these parameters.

Membrane conductance and cytoplasmic conductivity both fluctuated with a near 24-hour period ($p<0.001$ for peak values). A standard method in fitting circadian patterns is through a cosinor model. Membrane conductance fit a cosinor model of a 22.00 h period with an R-square value of 0.9113 while cytoplasmic conductivity best fit the cosinor curve at a period of 21.75 h with an R-square value of 0.9538. The near antiphasic appearance of the membrane capacitance and cytoplasmic conductivity suggests a cyclic ion movement from the cytoplasm across the cell membrane. No significant variation was observed in capacitance which is expected since there should be no membrane morphology changes across time points. These observations indicate an underlying cyclic activity of ion transport across the membrane.

In the absence of nuclei or other organelles in RBCs, these finding suggest membrane mechanisms involving differential ion channel activities and renders DEP a new technology to enhance the understanding of chronobiology. Future work will investigate the underlying mechanisms behind these changes and the effects of experimental conditions.

1.)O'NEILL, J. S. & REDDY, A. B. 2011. John S. Circadian clocks in human red blood cells. *Nature,* 469, 498–503

2.) HOETTGES, K. F., HÜBNER, Y., BROCHE, L. M., OGIN, S. L., KASS, G. E. N. & HUGHES, M. P. 2008.Dielectrophoresis-activated multiwell plate for label-free high-throughput drug assessment.*Analytical Chemistry,* 80, 2063-2068.

3.) IRIMAJIRI, A., HANAI, T. & INOUYE, A. 1979. Dielectric theory of multi-stratified shell-model with its application to a lymphoma cell. *Journal of Theoretical Biology,* 78, 251-269.

4.) FRY, C.H., SALVAGE, S.C., MANAZZA, A., DUPONT, E., LABEED, F. H., HUGHES, M. P. & JABR, R.I. 2012. Cytoplasm Resistivity of Mammalian Atrial Myocardium Determined by Dielectrophoresis and Impedance Methods. *Biophysical Journal,* 103(11), 2287–2294.

392098 Isoelectric Focusing: Current Limitations and Prospects

Abstract:

Isoelectric focusing (IEF) is an analytical and preparative technique which allows for analyzing amphoteric substances based on their isoelectric points (*pI*s). IEF plays an important role in protein characterization and proteomics related applications. Being discovered almost hundred years ago, IEF has gained its popularity in the sixties of the last century with the synthesis of complex amphoteric mixtures called carrier ampholytes (CAs) allowing for obtaining stable pH gradients with acceptable separation properties for different wide and narrow pH ranges. The most popular IEF applications, capillary electrophoresis and gel electrophoresis, rely essentially on CA technology. While the traditional formula for resolving power of IEF, already proposed by Kauman and Svensson more than fifty years ago, predicts resolution increase as a square root of voltage gradient, in a CA system the resolution is defined mostly by the "density" of CA preparations, a number of individual CA components per pH unit. Alternative approaches for pH gradient forming, in particular immobilized pH gradients (CA free) and thermal pH-gradients, are also analyzed and their resolving power is discussed.

392095 Surface-Enabled Isoelectric Focusing (sIEF) with Carrier Ampholyte Type pH Gradient

Abstract:

This works describes the first illustration of a newly developed, novel surface-enabled isoelectric focusing (sIEF) technique for micro-scale IEF applications. This new sIEF approach requires smaller applied voltages and eliminates the need for electrode reservoirs, which are required by traditional bulk gel IEF and macrochannel-IEF. Isoelectric focusing is capable of resolving proteins from complex mixtures based on small differences in their isoelectric point (pI). sIEF can be conducted at scales 100 times smaller than previously reported for macrochannel-IEF. Furthermore, the separation occurs in a gel printed on a surface, making it easily accessible for spot recovery and subsequent analysis.

To fabricate the sIEF gel, surface patterning tools (SPTs) available on a Bioforce Nano eNabler™ were utilized to print a 40 to 60 μm wide line of unpolymerized acrylamide solution spanning a 300 μm gap between micro-patterned, parallel gold electrodes on a traditional glass microscope slide. Broad range Pharmalyte™ carrier ampholyte solution (pH 3-10) was co-printed with the acrylamide monomer solution and then polymerized *in-situ*. Surface printing parameters were optimized with respect to carrier ampholyte concentration (4% wt/v), polymerization time (3h), surface properties of glass slide (UV-ozone treatment for hydrophobicity) and printing chamber humidity (15%). The established pH gradient was examined by pH-sensitive fluorescent dye FITC, with pH insensitive dye TRITC as intensity profile normalizer. Green fluorescent protein (pI ~6.0) and r-phycoerythrin (pI ~4.4) were then printed over the gel surface using the Nano eNabler™SPTs. DC voltages of 3V (100 V/cm) and 9V (300 V/cm) were applied across the 300 μm electrode gap to achieve protein IEF in the polymerized micro gel strip.

Our results describe the development of a 100 pL gel for surface-enabled IEF-PAGE. In addition, sIEF results illustrate optimization of the applied voltage, reduction of pH gradient drift, as well as both understanding and reducing the band broadening that occurs during focusing. In summary, this device operates in volumes 100 times smaller than previously reported macrochannel-IEF, and, requires less sample/reagent, relies upon a relatively quick fabrication process, and the surface is reusable. These attributes make it more attractive as a portable, user-friendly platform for widespread adoption and commercialization.

389587 Mathematical Modeling and Simulation Software for Electrophoresis

Abstract:

As this year Achievement Award goes to Prof. Pier Giorgio Righetti, his contribution to mathematical modeling, numerical simulation and software tools for electrophoresis is highlighted.

Despite the fact scientific fundamentals for calculating properties of aqueous ionic solutions in the range of molarities used in electrophoresis had been developed in the first half of the 20th century most of the electrophoresis practitioners in the second half of the same century used only simple Henderson–Hasselbalch equation or experimental methods for assessing pH of buffer solutions.

In the late 80s – early 90s Prof. Righetti and his co-workers developed algorithms and Windows application (Doctor pH) for optimizing mixtures of buffers and titrants for creating immobilized pH gradients and calculating pH of membranes. The efficient algorithms made it possible to run this application on the first versions of Windows on PCs of early 90s. A thorough experimental validation made it a useful tool for practical work.

Practical Capillary Electrophoresis research in Righetti's lab went together with a research on thermal heating inside the capillary and numerical simulations of electrophoresis in a capillary. Thermal studies resulted in creation of Capillary Zone Electrophoresis Assistant (CZEA) application used to determine the temperature of the buffer within the capillary during the electrophoresis run based on known parameters of the buffer and the capillary instrument or a calibration procedure.

The numerical simulations of the ionic transport served as an important link between the works of Milan Bier's group and modern electromigration modeling tools such as PeakMaster, Simul, Buffer Workshop, and SPRESSO.

Several of the currently available tools for modeling solution properties and elctromigration are reviewed and examples presented.

Joachim Kohn (1912-1987) and the Origin of Cellulose Acetate Electrophoresis

Richard M. Rocco, PhD
Samuel Merritt University
Oakland, CA 94609
November 19, 2014

Adapted from Rocco, R. M., Clin Chem 51(10):1896-1901 (2005) 1

Joachim Kohn (1912-1987)

Born in Poland in 1912
Received Medical Diploma, University of Lwow, 1936
Served in the Polish Army until 1947

2

Poland Invaded at the Start of WWII

The New York Times. EXTRA

GERMAN ARMY ATTACKS POLAND; CITIES BOMBED, PORT BLOCKADED; DANZIG IS ACCEPTED INTO REICH

The New York Times.

RUSSIANS DRIVE 40 MILES INTO POLAND, DEFENSE WEAK; POLISH PRESIDENT IN RUMANIA, WARSAW HOLDING OUT; BRITAIN AND FRANCE, CONCERNED, CONSULT ON ACTION

September 1, 1939 September 17, 1939

Germany and Russia divided Poland between them.

3

Katyn Forest, Russia

The Russians deported 15 000 captured Polish officers to three POW camps outside Smolensk, Russia. 4 254 Polish officers were executed in May 1940 and dumped into mass graves in the Katyn forest, 10 000 are still missing, 448 survived the executions (Raszeja, S., Forensic Sci Int 68:1-6,1994).

Joachim Kohn was one of the survivors.

4

US Congress Investigation Katyn Forest Massacre

82nd Congress 1952 | **J. Kohn, pg 538, Survived**

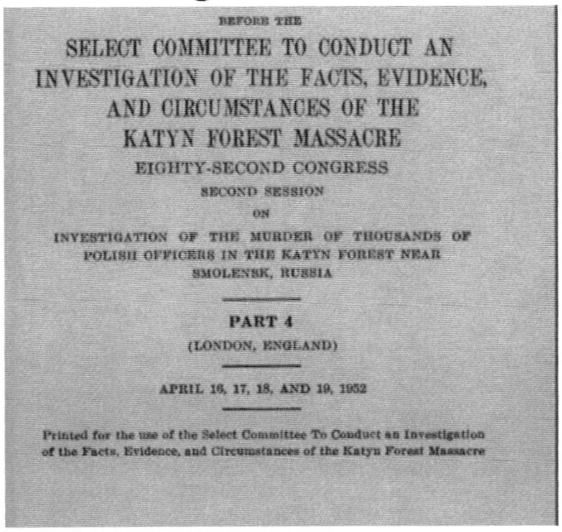
BEFORE THE
SELECT COMMITTEE TO CONDUCT AN INVESTIGATION OF THE FACTS, EVIDENCE, AND CIRCUMSTANCES OF THE KATYN FOREST MASSACRE
EIGHTY-SECOND CONGRESS
SECOND SESSION
ON
INVESTIGATION OF THE MURDER OF THOUSANDS OF POLISH OFFICERS IN THE KATYN FOREST NEAR SMOLENSK, RUSSIA

PART 4
(LONDON, ENGLAND)

APRIL 16, 17, 18, AND 19, 1952

Printed for the use of the Select Committee To Conduct an Investigation of the Facts, Evidence, and Circumstances of the Katyn Forest Massacre

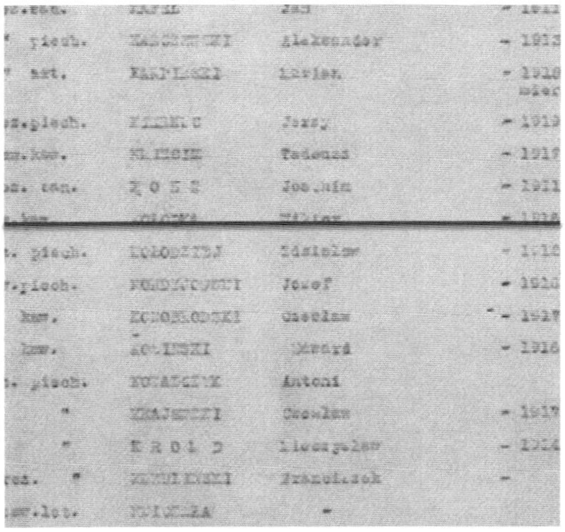

Joachim Kohn listed as a survivor in the 1952 Congressional Report.

5

Dr. J. Kohn, 1940-1950

- Made his way to Canada, joined the Polish Army in exile.
- Became a Commanding Officer of a field ambulance company, served in Italy and the Middle East.
- Received Military Cross and Silver Order of Merit with Swords.
- Decommissioned in 1947 and spent two years as a ship's surgeon.

6

Queen Mary's Hospital, Roehampton, London

Dr. Kohn joined Queen Mary's Hospital in 1950 as a pathologist in training. He was 38 years old.

7

Queen Mary's Hospital 1952-1953

Dr Kohn remained at Queen Mary's Hospital from 1950 until he retired in 1977 as Senior Pathologist.

All of his major discoveries in CA electrophoresis were accomplished while at Queen Mary's.

Dr. Kohn, sixth from the left, top row.

8

Paper Zone Electrophoresis

Filter paper was the most widely used support for serum protein electrophoresis in the 1950's. Discovered by Konig but made popular by Durrum and Tiselius starting in 1950.

P. König, 1937

(Actas e Trabalhos do Terceiro Congresso Sud-Americano de Chimica, Rio De Janeiro e Sao Poaulo 2:334-336)

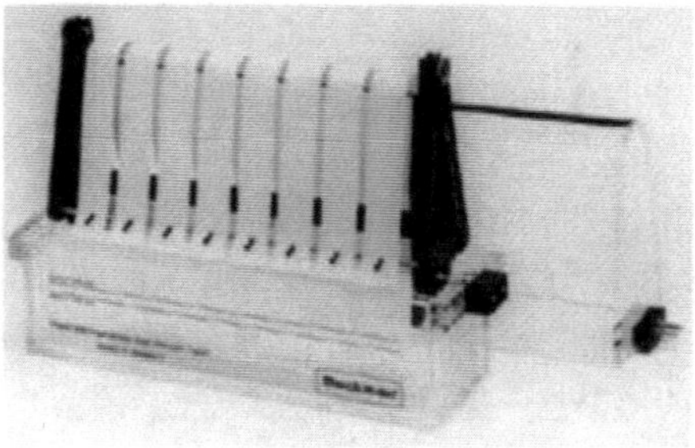

E. L. Durrum, 1950

(J Am Chem Soc 72:2943-2948)

H. -D. Cremer, A. Tiselius, 1950

(Biochemische Zeitschrift. 320:273-283)

9

The Problems with Zone Electrophoresis On Paper

- Long run times (6+ hours).
- High voltages with heat buildup and evaporation.
- Major tailing of albumin making quantitation a problem, other proteins like insulin bind and do not migrate.
- Poor separation of $alpha_1$ protein from albumin.
- Inability to make paper transparent for accurate densitometry.
- Poor protein detection limit with high sample volumes required.

10

E. L. Durrum Predicts a Better Support, 1955

"In the future probably, we will have better supporting media. Perhaps it will not be paper but some synthetic paper-like material".

The Future of the Technique in its Application to Clinical Research and Routine Analysis. pg 200, CIBA Foundation Symposium on Paper Electrophoresis, London, July 1955.

11

Fifty-Eight Years Ago This Week November 17, 1956

Fifteen months after Durrum called for a better support medium, Dr. Kohn introduced for the first time cellulose acetate (CA) for zone electrophoresis.

- 356th Meeting of the Biochemical Society
- Maudsley Hospital at the University of London

- ***A New Supporting Medium for Zone Electrophoresis***

J. Kohn (Introduced by E. J. King)

Biochemical J. 65:9P (1957) [abstract]

12

The Source of Innovation

- Dr. Kohn wrote a few years before his death,

"A chance remark by a commercial traveler that CA bacteriological filters could be made transparent by the use of a suitable clearing agent started a train of thought which led to the development of CA [for zone electrophoresis]".

Ritzmann, S. E., *Physiology of Immunoglobulins, Diagnostic and Clinical Aspects*, 1982, pg 15.

13

CA Membrane Filters in Microbiology

Goetz, A. in 1947 reported on the use of cellulose acetate filters in microbiology by Germany during WWII

(Slanetz, L.W. Applied Micro 3:46-51, 1955).

By the mid 1950's CA membrane filters were widely used to test for coliforms and other bacteria in water.

J Microbiology, ad 1955

14

CA Membrane Filters in Microbiology in Life Magazine

Life Magazine Aug 13, 1951

Collecting air samples Broughton St, Savannah, GA

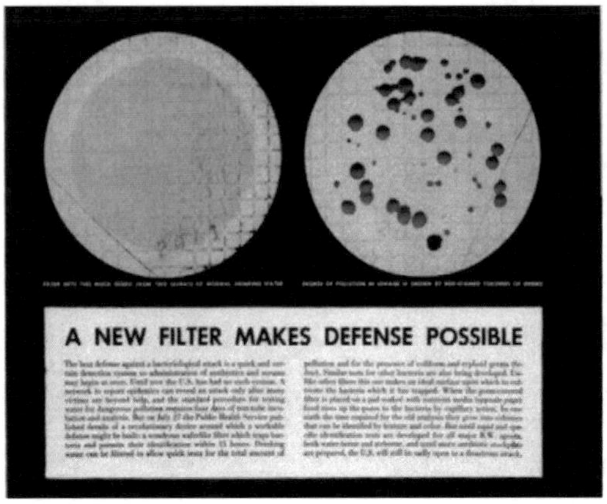
A NEW FILTER MAKES DEFENSE POSSIBLE

Life magazine article on the use of CA membranes to collect air born pathogens.

Making CA Transparent

- Numerous articles in the literature of the early 1950's described how to make CA membranes transparent.
- Following collection and staining of bacteria, CA filters can be made transparent with various immersion oils used in light microscopy that have a RI of > 1.450

- Richards, O. W.
  J. Bacteriol. 67(5):613 (1954)
- Ehrlich, R.
  J. Bacteriol. 70(3):265 (1955)

White Mineral Oil Used With CA Membranes

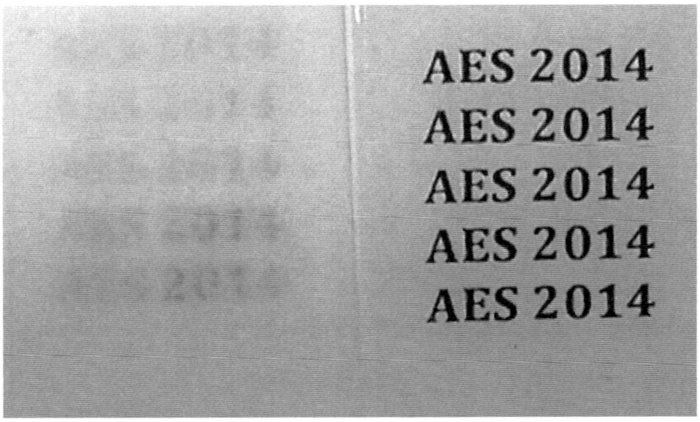

Whatman, 3MM Paper #3030 **Hoefer, CA # CE-M75**

Whatman filter paper or CA membrane wetted with Blandol® White Mineral Oil RI 1.466 (Sonneborn LLC) and placed over a white sheet of typed paper.

17

Serum Proteins on Paper vs CA

Wurm, M. Clin Chem 2:303-319 (1956)
Human serum 5.0 uL,
Whatman 3MM paper
61 cm long,
18 h separation
183 volts DC

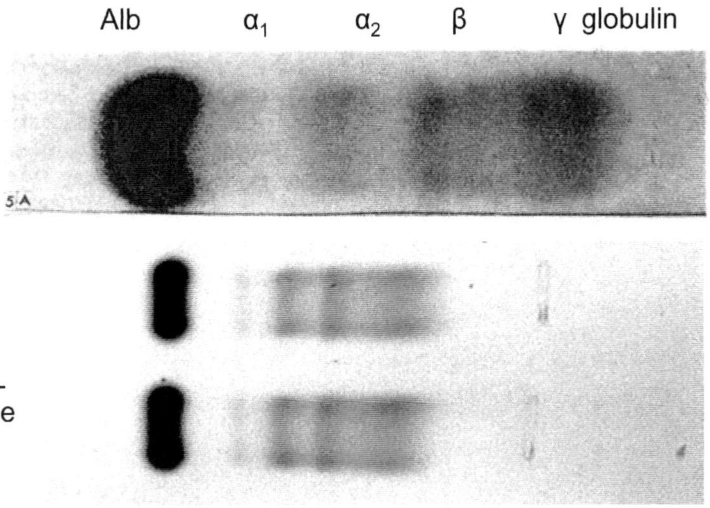

Human serum 0.5 uL
Hoefer CA Membrane
6 cm long,
30 min separation
150 volts DC

Human serum run on CA in the author's lab and stained with Coomassie Blue R250. Gray scale image.

18

CA Membrane Advantages

- Reduced protein tailing, sharper bands
- Made transparent with oils of RI of >1.474
- Dissolve in chloroform:ethanol (90:10) which made elution easier compared to paper.
- Barbital buffer pH 8.6, 0.05-0.1 ionic strength, with 30 min serum separations
- Easily detects < 5 ug protein with nigrosin.
- Insulin, lysozyme and histones separate well but not at all on paper.

19

5th Colloquium on *Protides of the Biological Fluids*

- Kohn's second paper on CA given May 6th, 1957 at St Jans Hospital, Bruges, Belgium
- Arne Tiselius (1902-1971) gives the opening address at the conference…..

SINT-JANSHOSPITAAL (SAINT JOHN'S HOSPITAL)

20

5th Protide Colloquium, Bruges May 6, 1957

Arne Tiselius in his opening address and round table discussion,

"..is it worthwhile to pursue paper electrophoresis to the bitter end? We all realize the danger of electrophoresis in the present form and that you can not evidently get results of clinical importance. Dr. Kohn has mentioned and described a substrate that seems very promising."

Peeters, H. (ed) *Protides of the Biological Fluids*, Proceedings of the Fifth Colloquium, Bruges, May 1957, pg 252.

21

5th Colloquium, 1957

- Kohn ends his presentation on CA zone electrophoresis with the following,
- *"I shall close this communication by expressing the hope that some of the readers will try this new medium…..I have no doubt that the greater experience and better facilities open to them will…help develop it into a real first-class tool for research."*

22

1957 -1958

- Kohn followed his presentation at the 5th Colloquium in May 1957 with a series of papers detailing the CA technique:

- June 29, 57 — Clin Chim Acta 2:297 (1957)
- Nov 15, 57 — 367th Biochemical Soc Meeting, Demonstration

- Jan 16, 58 — Nature 181:839 (1958)
- Mar 10, 58 — Clin Chim Acta 3:450 (1958)
- May 6, 58 — 6th Protide Colloquium, Bruges

23

The "Kohn" Electrophoresis Chamber 1958 (Clin Chim Acta 3:450)

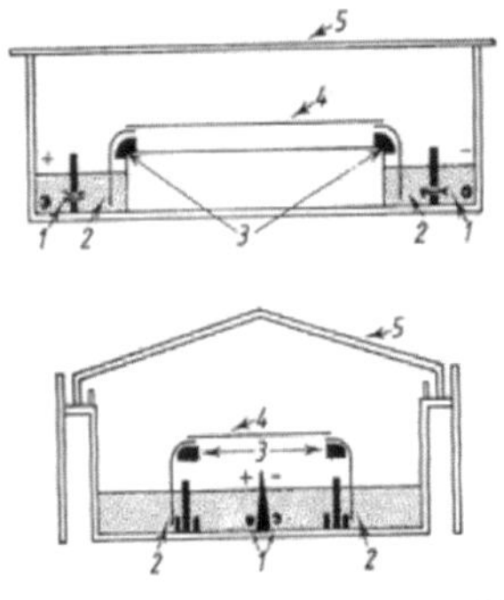

Fig. 2. Diagram illustrating the essential features of the old-style (top) and new-style (bottom) tanks: 1) electrode compartment; 2) buffer compartment; 3) adjustable bridge pieces; 4) CAM; 5) tank cover.

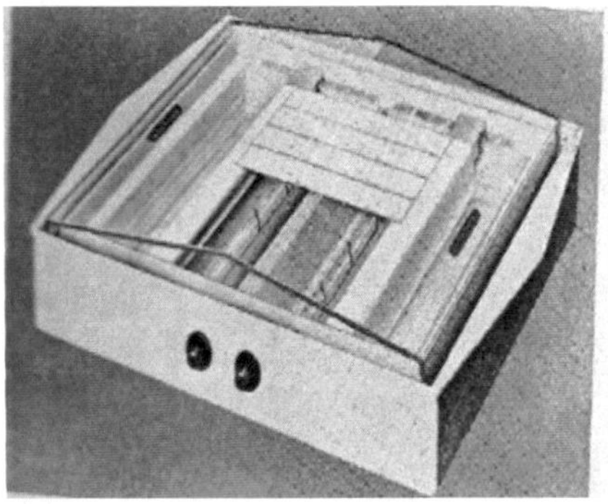

Kohn designs a new electrophoresis tank. The air gap in conventional paper tanks is eliminated, reducing both convection currents and the strip length. Dr. Kohn's design features were incorporated into tanks produced by Shandon, Gelman, Beckman, Helena, Hoefer etc.

24

The "Kohn" Chamber

- Shandon sells the "Kohn" Chamber but he states,
- "*The electrophoretic chamber [Shandon] described in this chapter has been developed and used for electrophoresis in my Department but this does not imply that other instruments and apparatus of a similar type are not capable of producing equally satisfactory results.*"

- Kohn, J. in *Chromatographic and Electrophoretic Techniques, Vol II Zone Electrophoresis*, Smith, I. (ed), Chap 5, pg 91 (1976).

25

Nigrosin (CI 50420),1958

- Kohn was a pioneer in the use of nigrosin protein stain following CA electrophoresis.

- Spore stain 1880's (Conn's Biological Stains, 9th ed, pg 392, 1991)
- First used by Ortega for proteins in paper . Nature 179:1087 (1957)
- Kohn adapts nigrosin for CA in 1958 and pioneers its use for low level protein staining over the next 30 years. Nature 181:839 (1958)

- Kohn's last paper published shortly after his death was on nigrosin Clin Chim Acta 166:335 (1987)

"***Nigrosin the forgotten protein stain***"

26

Nigrosin vs Coomassie R250 With BSA on CA Membranes

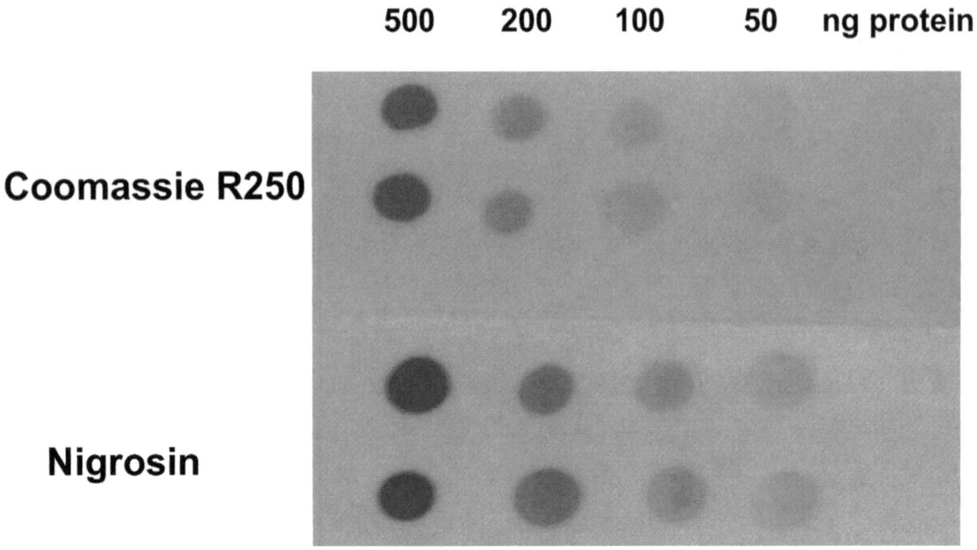

1.0 uL spots BSA on CA stained with either nigrosin or Coomassie R250 27

Nigrosin vs Coomassie R250 With BSA

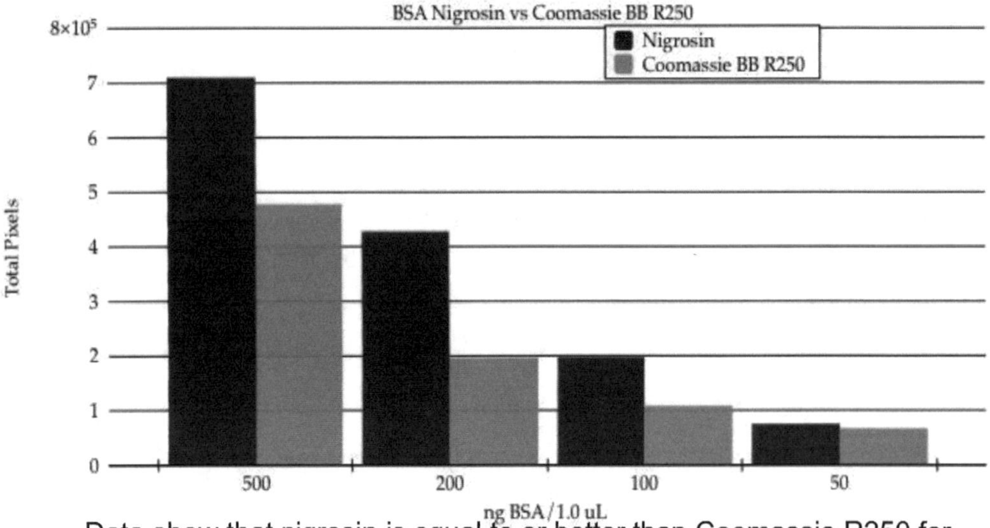

Data show that nigrosin is equal to or better than Coomassie R250 for staining BSA on CA membranes from 500 to 50 ng/1.0 uL per spot.

28

Nigrosin vs Coomassie R250 Gamma Globulin

500 200 100 50 ng

Coomassie R250

Nigrosin

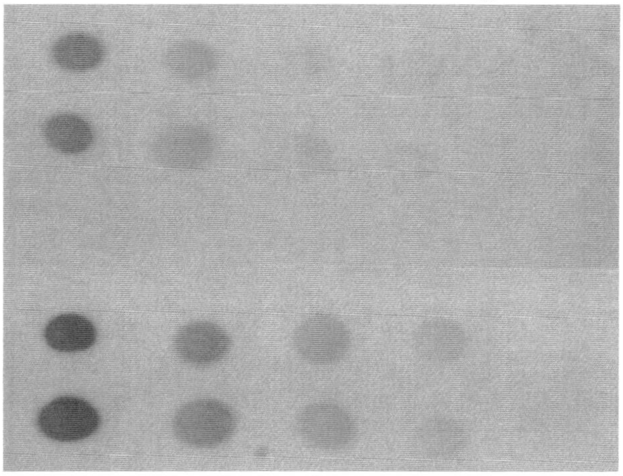

1.0 uL spots of bovine gamma globulin on CA membrane stained with nigrosin or Coomassie R250.

29

Nigrosin vs Coomassie R250 Gamma Globulin

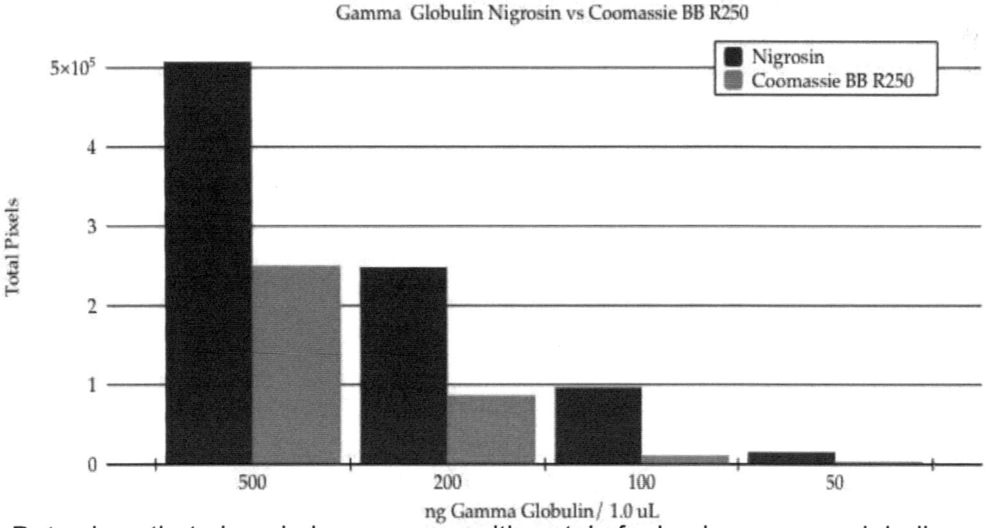

Data show that nigrosin is a more sensitive stain for bovine gamma globulin compared to Coomassie R250 from 500 to 50 ng/1.0 uL spots.

30

Hemoglobin Electrophoresis, 1958

- Kohn was the first to describe the separation of human hemoglobin phenotypes by CA electrophoresis.

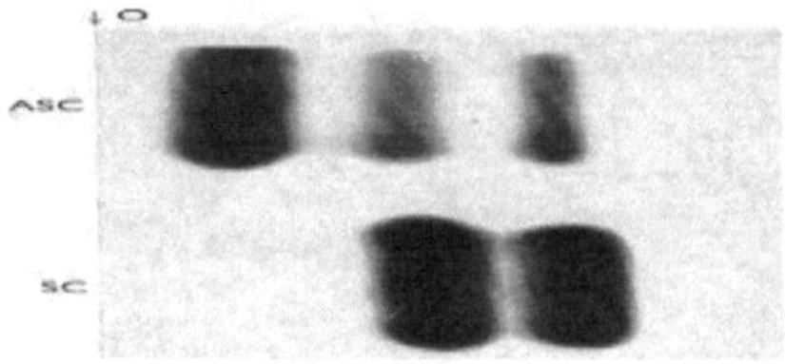

- Trans Royal Society Trop Med 52:304 (1958)
- Smith, I. Vol II Second Ed, pg 84-146 (1968)
- J Clin Path 22:109 (1969)
- Smith, I. Vol II Fourth Ed, pg 112 (1976) above

31

Labeled Substances,1960

Kohn demonstrated that radio labeled proteins and hormones separated well on CA membranes.

I^{131} Labeled albumin

I^{131} Thyroxine

Smith, I. Vol II, First Ed, pg 56-90 (1960)

32

Folic Acid Deficiency, 1961

- Kohn describes a rapid method for formominoglutamic acid (FIGLU) in urine for screening for folic acid deficiency.

  FIGLU is a precursor of glutamic acid that is elevated in the urine of patients with folic acid deficiency.

  Pyridine-acetic acid pH 5.3 buffer on CA followed by ninhydrin spray.

J Clin Path 14:345 (1961)
The Lancet 277:112 (1961)
J Clin Path 17:466 (1964)
Smith, I. Vol II Second Ed, pg 84-146 (1968)

33

Bioactive Substances, 1963

- Bradykinin and histamine from human urine detected by bioassay with the isolated rat duodenum after separation on CA membranes.

Brit J Exp Pathol 44:551 (1963)
Smith, I, Vol II Second Ed, pg 84-146 (1968)

34

Isoenzymes, 1968

- Some of the isoenzyme separations describe by Kohn on CA membranes.
- Lactic Dehydrogenase (EC 1.1.1.27)
- Malate Dehydrogenase (EC 1.1.1.37)
- Leucine Amino Peptidase (EC 3.4.11.1)
- Amylase (EC 3.2.1.1)

Brit Med J 285:170 (1982) Brit Med J 292:1523 (1986

Following the Chernobyl accident (April 26, 1986) Kohn suggests the use of a dip stick to detect salivary amylase to monitor for radiation exposure.

35

Kohn's Contributions to Smith's Vol II Zone Electrophoresis.

Smith, I. (Editor)
Chromatographic and Electrophoretic Techniques
Vol II Zone Electrophoresis

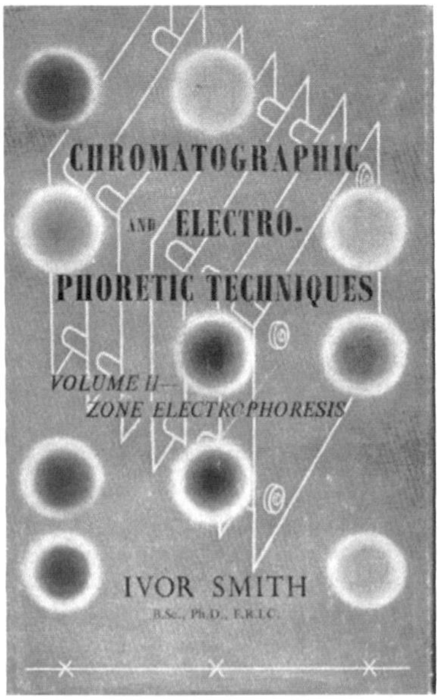

Kohn's chapters in the Smith books were procedure manuals for use of CA membranes.

Chapters on CA methods by Dr. Kohn
First Edition, 1960 pg 56-90
Second Edition, 1968 pg 84-146
Fourth Edition, 1976 pg 90-137

(no third edition published)

36

Staining Techniques

- Various staining techniques were adapted and improved by Dr Kohn for use with CA membranes.

- Nigrosin; ponceau S for proteins
- Periodic acid for glycoproteins
- Schiff's reagent/ozone for lipoproteins
- Leucomalachite green for hemoglobin

37

Kohn is issued two US patents for a multi-sample applicator for zone electrophoresis

United States Patent Office 3,505,858
Patented Apr. 14, 1970

APPLICATOR MEANS

ABSTRACT OF THE DISCLOSURE

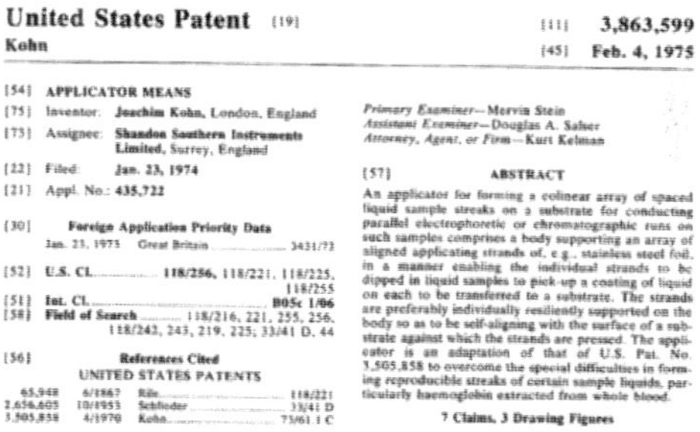
United States Patent [19] [11] **3,863,599**
Kohn [45] **Feb. 4, 1975**

[54] APPLICATOR MEANS
[75] Inventor: **Joachim Kohn,** London, England
[73] Assignee: **Shandon Southern Instruments Limited,** Surrey, England
[22] Filed: **Jan. 23, 1974**
[21] Appl. No.: **435,722**

[30] **Foreign Application Priority Data**
Jan. 23, 1973 Great Britain 3431/73

[52] **U.S. Cl.** **118/256,** 118/221, 118/225, 118/255
[51] **Int. Cl.** **B05c 1/06**
[58] **Field of Search** .... 118/216, 221, 255, 256, 118/242, 243, 219, 225; 33/41 D, 44

[56] **References Cited**
UNITED STATES PATENTS

65,948	6/1867	Rife	118/221
2,656,605	10/1953	Schlieder	33/41 D
3,505,858	4/1970	Kohn	73/61.1 C

Primary Examiner—Mervin Stein
Assistant Examiner—Douglas A. Salser
Attorney, Agent, or Firm—Kurt Kelman

[57] **ABSTRACT**

An applicator for forming a colinear array of spaced liquid sample streaks on a substrate for conducting parallel electrophoretic or chromatographic runs on such samples comprises a body supporting an array of aligned applicating strands of, e.g., stainless steel foil, in a manner enabling the individual strands to be dipped in liquid samples to pick-up a coating of liquid on each to be transferred to a substrate. The strands are preferably individually resiliently supported on the body so as to be self-aligning with the surface of a substrate against which the strands are pressed. The applicator is an adaptation of that of U.S. Pat. No. 3,505,858 to overcome the special difficulties in forming reproducible streaks of certain sample liquids, particularly haemoglobin extracted from whole blood.

7 Claims, 3 Drawing Figures

April 14, 1970 J. KOHN **3,505,858**
APPLICATOR MEANS
Filed Nov. 16, 1967 3 Sheets-Sheet 1

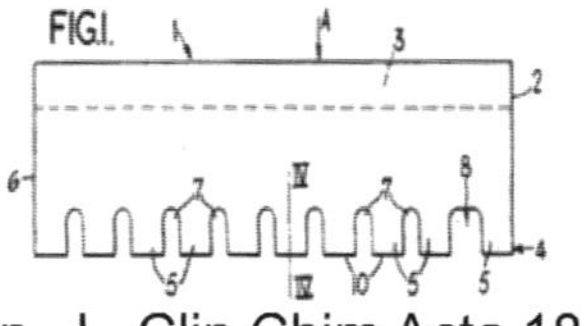

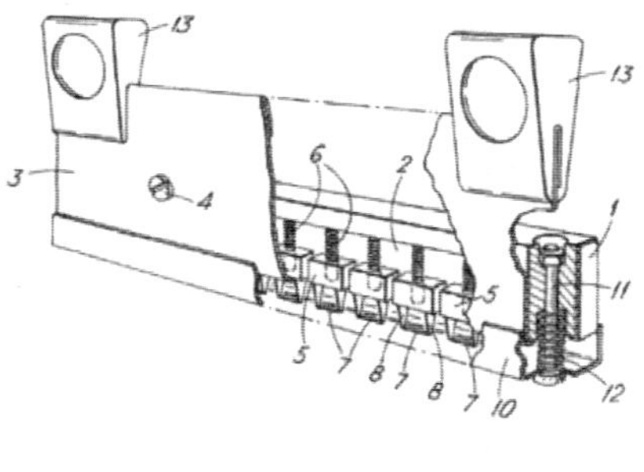

Kohn, J., Clin Chim Acta 18:65 (1967)

Dr. Kohn's Stamp Applicator

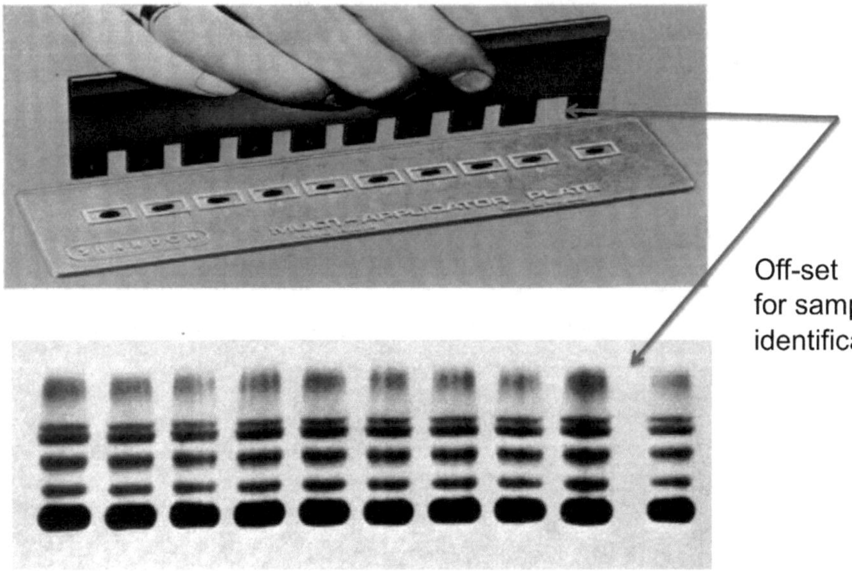

Off-set
for sample
identification.

39

Immunoelectrophoresis,1957

[Grabar, P., Williams, C.A. Biochim Biophys Acta 17:67 (1955)]

As early as 1957 Kohn adapts the Grabar & Williams technique to CA.

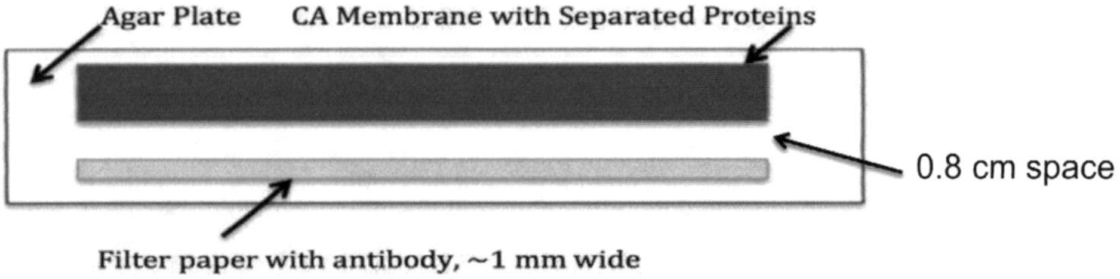

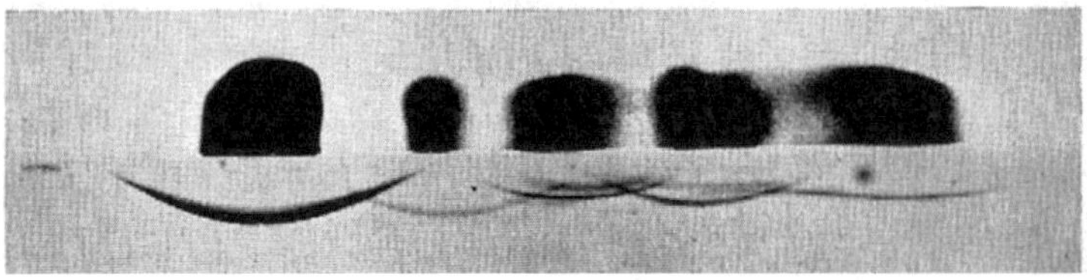

From, Kohn, J. Nature 180:986 (1957)

40

Immunoelectrophoresis

[Grabar, P., Williams, C.A. Biochim Biophys Acta 17:67 (1955)]

Kohn continued to refine and improve on the Grabar & Williams immunoelectrophoresis technique with CA membranes.

- Clin Chim Acta 3:450 (1958)
- Smith, I. Vol II First Ed, pg 56-90 (1960)
- Smith, I. Vol II Second Ed, pg 84-146 (1968)

41

Double Immunodiffusion, 1959

[Ouchterlony, O., Acta Path Micro Scand 26:507 (1949)]

Kohn replaces agar with CA membranes for Ouchterlony's double immunodiffusion technique with serum samples surrounding a central antibody sample.

- Nature 183:1512 (1959)
- Smith, I. Vol II, First Ed, pg 56-90 (1960)
- Smith, I. Vol II, Second Ed, pg 84-146 (1968)
- Nature 217:1261 (1968)

42

Bidimensional Ouchterlony-Elek Method On CA

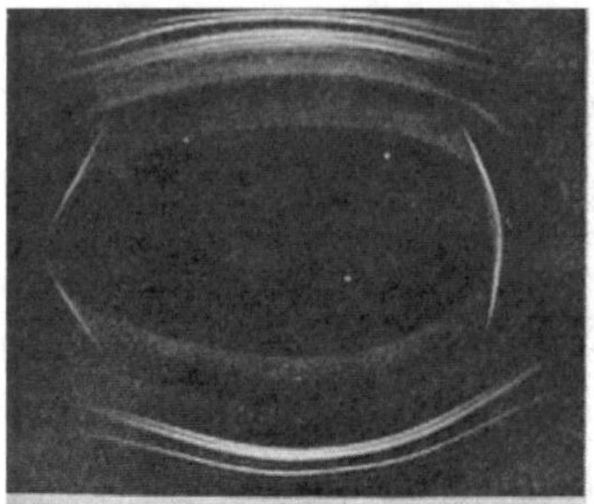

Horse antiserum in the center of CA membranes with human serum on the periphery.

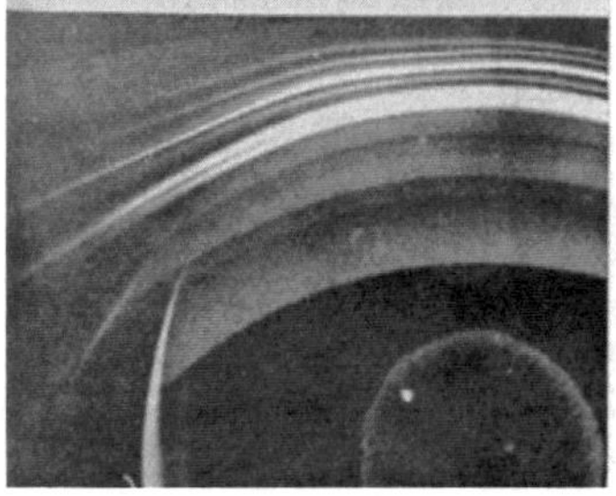

Enlarged image of the precipitation patterns on CA membranes, cleared and stained with nigrosin.

Smith, I. Vol II pg 130-131 (1976)

Double Immunodiffusion, 1959

- Kohn used CA with the immunodiffusion technique to measure plasma C-reactive protein (CRP) in plasma and showed that elevated levels correlated with the severity of the patient's burns.

- Lancet 275:334 (1960)
- 8th Protides Colloquium pg 315-318 (1960)

"Roehampton Burns Dressing" 1963

- Following the work with CRP on CA Kohn designs a new burn dressing made from polyurethane foam that is adopted by the Royal Army Medical Corps as standard issue.

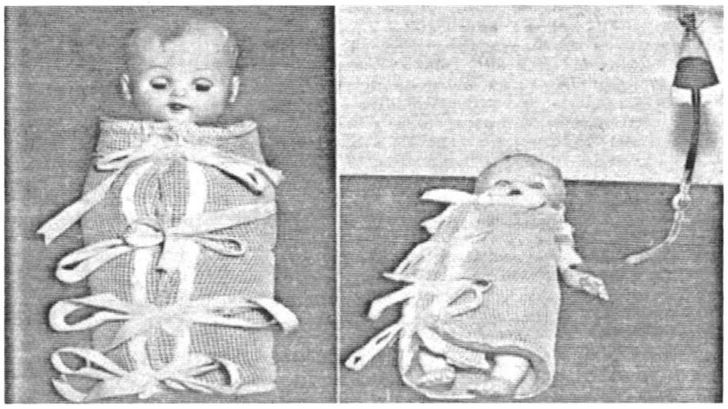

- Journal Royal Army Medical Corps
  *A transport dressing for extensive burns.*109(4):199-202 (1963)

45

Single Radial Immunodiffusion, 1968

[Mancini, G. et al. Immunochemistry 2:235 (1965)]

- Kohn adopts the Mancini technique for CA membranes
- He replaces the agar with squares of CA impregnated with antibody for serum IgG, IgM, IgA analysis.

- Smith, I., Vol II Second Ed,, pg 134-135 (1968).

46

Immunochromatography From Sephadex® 1968

[Agostoni, A. et al., J Lab Clin Med 69:522 (1967)]

Proteins separated on Sephadex were blotted onto CA membranes for immuno detection.

- Smith, I. Vol II Second Ed, pg 84-146 (1968)
- Immunology 15:863 (1968)
- Smith, I. Vol II Fourth Ed, pg 90-137 (1976)

47

Counter Immunoelectrophoresis, 1968

[Bussard, A. Biochim Biophys Acta 34:258 (1959)]
Kohn developed CA membrane methods for the detection of a wide variety of proteins by counter immunoelectrophoresis.

<u>Hashimoto's Disease</u>

- Smith, I. Vol II Second Ed, pg 84-146 (1968)

<u>Australia Antigen</u>

- J Immunol Methods 11:303 (1976)

<u>Alpha Fetal Protein (AFP)</u>

- Lancet 304:334 (1974)
- Smith, I. Vol II Fourth Ed, pg 90-137 (1976)

48

Counter Immunoelectrophoresis (CEP),

- An early application of CEP according to Bussard but CA membranes for HBAg stained with nigrosin.

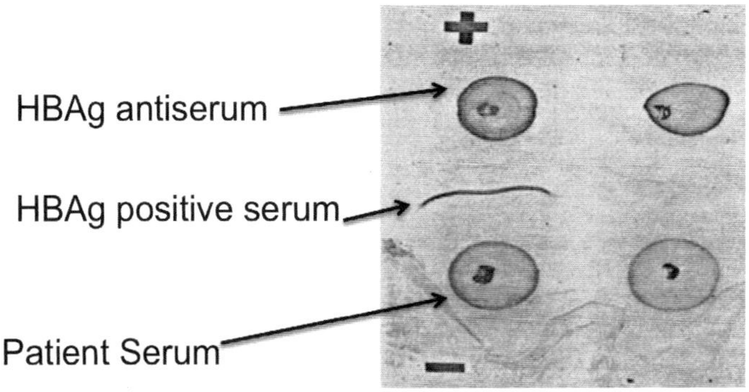

Kohn, J., J Immunol Methods 11:303-309 (1976)

49

Awards and Honors (partial list)

- Abrogina Medal, burn treatment (Italy)
- Hirai Medal, cancer detection (Japan)
- Kotama Medal, electrophoresis (Japan)
- Honorary Member, Assoc Clin Biochemists
- Director of the Guildhay Protein Unit
- Kohn Memorial Lecture, Royal College of Pathologists
- Founding Member, Int Soc Oncology and Biomarkers

50

- Dr. Joachim Kohn died on March 31, 1987 at age 75 and was survived by his wife Wiera, daughter Ewa and brother Alexander.
- "*he had a strong character tempered by kindness and genuine modesty.*"

British Med J 294(6579):1105 (1987).

51

Special Thanks To:

Linda Lamb
Gordon Jones
Queen Mary's Hospital
Archive & Museum Group

Dr. Vincent Marks
University of Surrey
Guildford, Surrey, UK

Dr. Pier Giorgio Righetti
Politecnico di Milano
Milano, Italy

52

Thank you

Richard M. Rocco, PhD

rrocco@samuelmerritt.edu

53

362912 The Monkey King: A Personal View of the Long Journey Towards a Proteomic Nirvana

Abstract:

The review covers about fifty years of progress in "proteome" analysis, starting from primitive two-dimensional (2D) maps attempts in the early sixties of last century. The polar star in 2D mapping arose in 1975 with the classical paper by O'Farrell in J Biol. Chem. It became the compass for all proteome navigators. Perfection only came, though, with the introduction of immobilized pH gradients, which fixed the polypeptides spots in the 2D plane. Great impulse in proteome analysis came by introducing informatics tools and creating databases, among which Swiss Prot remains the site of excellence. Towards the end of the nineties, 2D chromatography, epitomized by coupling strong cation exchangers with C_{18} resins, began to be a serious challenge to electrophoretic 2D mapping, although up to the present both techniques are still much in vogue and appear to give complementary results. Yet the migration of "proteomics" into the third millennium was only made possible by mass spectrometry (MS), which today represents the standard analytical tool in any lab dealing with proteomic analysis. Another major improvement has been the introduction of combinatorial peptide ligand libraries (CPLL), which, when properly used, enhance the visibility of low-abundance species by 3 to 4 orders of magnitude. Coupling MS to CPLLs permits to explore at least 8 orders of magnitude in dynamic range on any proteome.

AUTHOR INDEX

AUTHOR INDEX

AIChE
120 Wall Street, FL 23
New York, NY 10005-4020

ISBN 978-1-5108-1243-7

CURRAN ASSOCIATES INC.
proceedings.com